Animal Suffering and the Problem of Evil

Animal Suffering and the Problem of Evil

NICOLA HOGGARD CREEGAN

Oxford University Press is a department of the University of Oxford.
It furthers the University's objective of excellence in research, scholarship, and education
by publishing worldwide.

Oxford   New York
Auckland   Cape Town   Dar es Salaam   Hong Kong   Karachi
Kuala Lumpur   Madrid   Melbourne   Mexico City   Nairobi
New Delhi   Shanghai   Taipei   Toronto

With offices in
Argentina   Austria   Brazil   Chile   Czech Republic   France   Greece
Guatemala   Hungary   Italy   Japan   Poland   Portugal   Singapore
South Korea   Switzerland   Thailand   Turkey   Ukraine   Vietnam

Oxford is a registered trademark of Oxford University Press in the UK and
certain other countries.

Published in the United States of America by
Oxford University Press
198 Madison Avenue, New York, NY 10016

© Oxford University Press 2013

New Revised Standard Version Bible, copyright 1989,
Division of Christian Education of the National Council of
the Churches of Christ in the United States of America.
Used by permission.
"Theodicy" (9.1) from UNATTAINABLE EARTH by CZESLAW MILOSZ and
TRANSLATED BY ROBERT HASS. Copyright © 1986 by Czeslaw Milosz and
Robert Hass. Reprinted by permission of HarperCollins Publishers.

All rights reserved. No part of this publication may be reproduced, stored in a
retrieval system, or transmitted, in any form or by any means, without the prior
permission in writing of Oxford University Press, or as expressly permitted by law,
by license, or under terms agreed with the appropriate reproduction rights organization. Inquiries
concerning reproduction outside the scope of the above should be sent to the
Rights Department, Oxford University Press, at the address above.

You must not circulate this work in any other form
and you must impose this same condition on any acquirer.

Library of Congress Cataloging-in-Publication Data
Hoggard Creegan, Nicola.
Animal suffering and the problem of evil / Nicola Hoggard Creegan.
p. cm.
Includes bibliographical references (p.) and index.
ISBN 978–0–19–993184–2 (hardcover : alk. paper)
1. Animals—Religious aspects—Christianity.   2. Animal welfare—Religious
aspects—Christianity.   3. Good and evil—Biblical teaching.   4. Evolution
(Biology)   5. Tares (Parable)   I. Title.
BT746.C73 2013
231'.8—dc23
2012037743

ISBN 978–0–19–993184–2

1 3 5 7 9 8 6 4 2
Printed in the United States of America
on acid-free paper

For Charlie

Contents

Acknowledgments	ix
Introduction	1
1. Animals in the Garden of Eden	14
2. Humans, Animals, and Death Revisited	27
3. Animal Suffering—Philosophical Responses	44
4. Animal Suffering—Theological Responses	56
5. The Best of All Possible Worlds?	71
6. The Wheat and the Tares: Re-Imagining Nature	82
7. A Picture Held Us Captive	97
8. New Dynamics in Evolutionary Theory	110
9. Dualism or Tares in Evolutionary History?	127
10. The Fall and Beyond	138
11. Concluding Ethical Reflections	154
Epilogue	173
Notes	177
Index	201

Acknowledgments

THIS BOOK HAS been in some ways a lifetime in its coming, and at least five years in its writing. I acknowledge therefore, my first theologian, Pierre Teilhard de Chardin. He transfixed me and set me off on this journey at the young age of 12. I am grateful to friends and students and family far and near for God conversations on issues of evolution, humanity, freedom, and our connections to the animal world.

I am very grateful to Cynthia Read and OUP for enthusiastically taking up this project when it was all but written and providing helpful guidance and advice along the way.

My deep thanks to Christopher Southgate for coming to New Zealand in 2009 to talk about the *Groaning,* and thereby furthering all our thinking in this area. Thanks to Christopher and Andrew for the heartening conversation on these issues in different parts of the world.

I must acknowledge the Templeton Foundation for a grant used in the early stages of this project, and also the encouragement and wide input of the many Templeton funded Science and Religion conferences in the years 1998 and following in Oxford, Boston, Philadelphia, Raleigh, Canberra, and Brisbane. I benefited also from Templeton course program awards and from the funding that has come through Metanexus and its conferences in Phoenix and Philadelphia.

I am particularly grateful to the Templeton Foundation and the CCCU for the Summer Seminars in Science and Christian Faith in the summers (Northern) of 2003–2005 at Wycliffe College in Oxford, and for all the inspiring and splendid participants, especially the Isis frogs and leaders John Roche, Alister McGrath, and the late Ernan McMullan. I remember especially the kindness and wisdom of Ernan. The conversation some of us had with him under the tree on the lawn of Wycliffe College one summer afternoon was, I thought at the time, a foretaste of heaven. As one who has tried to juggle teaching, cross-disciplinary research, and parenting I would never have been able to tackle these questions without this stimulation and input, though I take full responsibility for the particular stance I present here.

Thanks also to those who have read portions of the manuscript—Judith Brown, Rod Thompson, Tim Meadowcroft, Graham O'Brien, Charles Creegan, Christine Pohl, and reviewers. A special thanks to Phil Church for bringing his biblical and editorial eye to the whole manuscript. The mistakes and idiosyncrasies are all mine.

Thanks to Mon Subritzky and Joshua Schoombie (my teaching assistant in 2012) for checking references and for your enthusiasm, and to Mon for finding the Polish poet Czeslaw Milosz.

I am grateful also to Steve and Cathy Ross and family, for your long friendship and for putting up with me for extended periods in Oxford in at least four different summers, and to Margaret Yee for arranging Bodleian access and a visiting fellowship in the summer of 2007.

Thanks to Ruth Barton who came to the beach as writing companion in the years 2009–2012. The Pacific is a magnificent backdrop for all paltry attempts to understand nature. Thanks also to Milu, now sadly deceased, for walks on the beach and for the much needed introduction to the life of a dog.

Large sections of this book were completed during leave from Laidlaw College in 2009, for which I am grateful. Thank you to Laidlaw students who have interacted with me on all issues theological and evolutionary in the last decade.

Last but not least I am thankful to my family, especially Charles, who has held the fort while I traipsed off to other parts of the world, who read it all, and helped with all the complexities of indexing.

I finished the first last draft of this book in the week of the Christchurch Earthquake, February 22, 2011. This earthquake, the most sudden and destructive in a long series of aftershocks of a previous earthquake some six months before, destroyed large parts of the old Christchurch city. One hundred and eighty nine people lost their lives in a moment in liquefied earth and falling masonry. The cathedral at the center of the city was destroyed. The aftershocks in Christchurch have continued unabated ever since. It is a reminder that sudden catastrophe strikes as often close to home as it is does in times and places distant from us. It is a reminder of the fragility of the natural world, and of the way in which life is lived always and inexplicably close to the edge of chaos.

Piha, May, 2012

*Animal Suffering and
the Problem of Evil*

Introduction

Then the Lord God said, "See, the man has become like one of us, knowing good and evil."
Gen 3:22

FOR THOSE OF us coming to faith in the second half of the twentieth century theology has always been done "after the Holocaust," the great eruption in common life, but one that is symbolic also for all the other massacres in the last hundred years. Sixty years later, contemplation of the *Shoah* is only now distant enough to be reflected in novel, film, and drama. What peace there has been in the last fifty years was lived in the shadow of possible nuclear holocaust and with the memory of Hiroshima and Nagasaki, and in the turmoil of Vietnam, Cambodia, India and Pakistan, Afghanistan, Timor, Korea, Uganda, Rwanda, the Congo, Bosnia, and the ever simmering Middle East with all their traces of postcolonial strife. Twentieth century theology agonized over all of this at the same time that we lost all confidence in the one major theodicy of Western theology—the Adamic Fall.

Silence is and was always one response to evil. Any attempt at explanation only trivialized that which we attempted to defuse. In the end one must speak. In the context of such evil the response that most typifies the twentieth century theology was that of Jürgen Moltmann. His theology of hope was also his personal answer to despair: God meets us in solidarity with God's own suffering, epitomized by Jesus as the Son of God, dying on a cross. Whether one thinks this satisfactory or not, it is one way of beginning to live with God after the Holocaust.¹

Secular responses are also hopeful and focus upon a determination to do better, especially collectively and corporately. It is here that the problem of animal suffering emerges. For humans are also animals, though for much of the modern age we have satisfied ourselves that other animals were sufficiently unlike us that we did not have to agonize over their suffering, nor our own animal natures. Darwin began to focus our attention on our animal origins; yet only in the last decades of the twentieth century was it evident that high degrees of sentience and almost certainly of suffering are possible for creatures. Moreover, this suffering must have occurred for many millennia before humanity emerged; human proclivities for violence were anticipated in the predatory and inquisitory skills of our more innocent animal ancestors. There is no physical paradise to which we can return; none we can remember, though the symbol of Eden as the true locus of human happiness continues to inspire and motivate us. Does the deep history of human violence, together with our discoveries that so much of what we do is motivated by deep drives beyond our consciousness, call into question our attempts to do better?

These two responses to evil turn out to be insufficient. Moltmann brought to light a neglected aspect of atonement and helped extend the boundaries of salvation to the wider cosmos. The problem of why it all had to be this way remains. Why must life be lived in such close proximity to horror? The gazelle caught in the web of deliberate predation by a lion has no knowledge of God's solidarity, if indeed that solidarity is there for the animal, as well as for us. The second response leads us into a quandary. How is doing better defined if humans are in fact animals and God has guided or allowed the evolutionary process; are we then meant to be vicious and predatory? Is God really a loving God? Does God care about animals who suffered for many eons before humans came to be? By what power can humans escape their deep-seated drives? Why does God do it all this way, when in the end—in Christ God reveals Godself to be loving and to demand love?

Which God?

Philosophers also grapple with the omni-God, the God who, if God is God, must be all powerful, all knowing, and perfect in love. In the twentieth century the temptation has been to solve the problem of the omni-God by denying that this is God, as is the case in both process theology and deism. Even more orthodox biblical theologians will say that we should not suppose that the God of the Scriptures *is* the omni-God. Tom Wright, for instance, says of the early Jewish believers:

> Emboldened by deep-rooted traditions, they explored what appears to us a strange, swirling sense of a rhythm of mutual relations within the very being of the one God: a to-and-fro, a give-and-take, a command-and-obey, a sense of love poured out and love received.[2]

This God is not an omni-dictator God; rather God *is* mutuality and dynamic interacting love. Wright argues that we have taken some of the many metaphors of God and hardened them into a philosophical absolute. Michael Welker also speaks of the subtle difference between the "production God" of popular understanding and the enormously subtle creator God of Genesis:

> I have been greatly surprised to discover major divergence between, on the one hand, the concepts of creation implicit in Genesis 1 and 2, which are normative creation texts for the Jewish-Christian traditions, and on the other hand, the *concepts* of creation that are dominant in Western cultures.[3]

He speaks of the "paltry conception of an ultimate process of causing and being caused," and of a "post-mechanistic" concept, which connects "images of production and of the exercise of power." In contrast to these common views of a causal God Welker claims that

> Genesis 1 and 2 describe the entire creation as in many respects having its own activity, as being itself productive, as being itself causative. *The creature's own activity, which is itself a process of production, is not only a consequence and result of a creation that is already completed. Rather it is embedded in the process of creation and participates in that process.* The Priestly creation accounts describe *the whole creation*, not only and not firstly human being, as itself active, separating, ruling, and imparting rhythm, as itself producing and giving life.[4]

Creaturely creativity and the "bringing forth" of the natural world is, of course, what we observe, what we hear is going on in the evolutionary process. This fits better with evolution than does the production model of popular imagination. Which God do we choose?

Whichever God we choose we end up bumping up against the puzzle of evil. An interacting God might be expected to have fashioned better results, like an attentive, good parent might produce, for instance. This is a problem not only for our minds—our philosophy and theology— but also for our experience of God. For if theologies of God are manifold so also are forms of religious *experience*. Many experience God as remote and unconnected to ordinary life; indeed they hope in their suffering that God is *not* closely involved. Others experience or have experienced a deep visionary sense of being at one with the universe, or have experienced God as the charismatic provider of healing and deliverance. At one end of the spectrum Charismatic Christianity can become triumphalistic. At the other end, theology after the *Shoah* appears to have lost the expectation of God's intimate presence, to have lost belief in prayer.

This loss of confidence, even within the church, may in fact know no bounds. How does the church reconcile these seemingly opposed but authentic responses of worship and lament?

An often neglected aspect in this dialogue, and in our response to God in twentieth century theology, is the charismatic revivals that largely refilled the churches through the latter half of the century and made explicit the connection between Christ and Spirit. Yet charismatic revival has confidence that God acts, God answers prayer, God heals, albeit in ways we do not understand. The stories of these healings and those from every previous generation that speak of God's love and faithfulness and provision—stories that live within the paradigm of gospel healing, miraculous provision, and deliverance from peril—have about them the ring of authenticity and are consistent with the dynamic interacting God described earlier. Yet how do we live with a God who provides and does nothing at the same time? We tend to shut out either one divine aspect or the other. Hence fundamentalism, on the one hand, minimizes the problem of evil. Liberalism, on the other hand, combines naturalism and a denial of the miraculous.

Into this already fraught context comes one more problem, that of the brutality of evolution and animal suffering. The twentieth century was a time when the deep past came clearly into our sights and our consciousness. The last decades of the twentieth century brought into focus the four billion years of life on earth, which had never much been pondered. With this history came the shocking realization that the course of evolution, a process that has taken up most of deep time, was filled with death and suffering and extinction. Animals seem to have suffered. Animals perhaps knowingly inflict suffering. This is one more reason not to believe in the omni-God, nor in the God of the Scriptures. How could God have brought humanity to be by this long, long process of ordeal only to turn around and say to us that love is the central law of the universe—not survival of the fittest, not competition to the death, not nature red in tooth and claw—but love. It is not surprising that the full impact of all of this on thoughtful and sensitive people is often a loss of faith or, in many cases now, the complete lack of any understanding of why anyone would believe.

This topic, of inordinate suffering which began long before humans emerged, has been the subject of many recent and thoughtful books. Why another book? Surely this is only proof that we do not know. My only excuse is that there is still much searching to do. If we don't keep talking the metaphysical high ground is taken by Richard Dawkins and others like him. In this book I will make no attempt to cover the exhaustive range of materials included in previous books, especially Michael Murray's *Nature Red in Tooth and Claw* and Christopher Southgate's *The Groaning of Creation*. I hope to bring to this conversation a different way of looking at the experience of being a Christian and a different perspective on what is happening in the evolutionary process. We cannot exclude from our experience either the God who answers prayer and produces miracles or the God who is absent and aloof; both light and darkness are our witness. Nor can we exclude from our experience and knowledge of the natural world either its cruelty or its perfection. Both are present and must be taken into consideration. I will emphasize, then, more than others have done the sense of encounter between light and darkness that the gospel world evokes. And I take into consideration the new vistas for rapprochement with theology that recent developments in evolutionary theory proffer. In the last decade or two the way in which we understand the evolutionary process has changed and is changing. There is an opening up of evolution to new hermeneutics that are more compatible with theological understandings.

We must also continue this exploration *because animals need us*. As human animals we have yet to understand what dominion means, and we have only just begun to think widely about the ethics of eating meat and animal experimentation. We *also need animals*. Only if we do manage to reconnect with nature, at a time when collectively we are looking for technical solutions to our ecological crises, might we again trust our deep religious sense, or what John Calvin calls our *divinitatis sensum*, and trust God in spite of evil.[5]

Moreover as humans we must solve the problem of violence. Violence is close to the heart of our founding myths, and it has been deeply embedded in our theology. The evolutionary process has been violent—as well as cooperative. How do we read this ancient history? Is it a model or does the violence need resisting? I am writing, then, on the edge of a revolution in human thinking and especially in theology. The sentience and importance of animals and our close links to them will unsettle and disturb the parochial and anthropocentric theology of the modern age.

The Argument of This Book

My response, which I will develop in this book, is that in spite of the evil that has been made evident in the evolutionary process and the suffering of animals and in spite of the seeming absence of God—in evolution, in piety, in history, in the *Shoah—we can believe in the goodness of God if we have other reasons to believe that God is good*. This is the answer that comes from the book of Job or Lamentations—not an explanation for evil, but an affirmation of God's love, an affirmation that is easily undercut by current atheistic arguments by the likes of Richard Dawkins, who flaunts nature's imperfections as evidence of the incoherence of theism.

How do we come to acknowledge that God is good in spite of evil? This knowledge is only made possible by a certain *reading* of nature. Is it a chaotic mixture? Is it essentially purposeless? Does it bear the marks of goodness and also beauty, or does the shadow side of nature invalidate its qualities? I argue that nature is indeed shot through with beauty and perfection. We might witness this beauty in the dawn of a new day in a remote wilderness area or in an awareness of life's extraordinary complexity and beauty. Two things might undermine this "right brain" recognition.[6] The first is the sense that although we *feel* this way it is all just the result of blind chance; such recognition undermines our experience. The second is the knowledge that if we dig deeper we will find disease and suffering and predation and precarious lives lived on the edge of survival. I will argue, however, that if we trust our instincts and experience we can view nature as both wheat and tares, the tares not eclipsing or cancelling out the good, but rather possibly even upholding the good.

I argue, then, that we should apply the parable of the wheat and the tares outward from the domain of church and society, into the creation from which the parable comes. In the parable of the wheat and the tares (Matt 13) Jesus speaks of a field filled with good wheat and entangled weeds, or tares. The workers want to root out the tares—which have been sowed in the good field by the "enemy," but Jesus cautions that this way wheat will also be uprooted. Better, he urges, to let both grow until the day of judgment when separation will occur and the tares will be burnt. Jesus is arguing from the manifestly obvious and well-known situation that applies in the field of wheat to the human sphere. I wish to argue in reverse that laws or metaphors like that of wheat and tares apply also to the whole biological sphere.

First, affirming a "wheat and tares" universe allows us to see the world not as a murky set of imperfections and the reasonably good, but as good and evil deeply entwined for all of the history of life. In this way both good and evil can be seen and recognized without having to baptize what is patently not good, nor explain away the goodness that has always spoken to us of the presence of God, nor ultimately even decide what is good and what is evil because they are so closely entwined—and may even change their value depending on their circumstances and time period. A wheat and tares universe is revealed partly in Old Testament lament, in which the evil is acknowledged but is understood in the long context of God's ultimate involvement.

Second, a wheat and tares approach takes note not just of our experience of nature, but also recent evolutionary theories. We are called to take seriously, as well, newer and broader and emergent renderings of evolutionary theory that articulate natural selection and "nature red in tooth and claw" as only one description of the evolutionary process, which is otherwise characterized by surprising convergence and cooperation. In animals we see both our failures and our greatness in latent form. I argue that all of this requires that theology must recognize the deep embeddedness of humans within nature, within the animal kingdom, and recognize the continuities as well as the discontinuities between ourselves and non-human animals.[7] In doing this we open up new vistas for theology in terms of the possible salvation of animals, taking seriously Isaiah 11:6–9, and 65:25[8] and rethinking dominion; how can we improve animals' lives rather than increasing their hardship? This view would assume that communication and care across species is intended and proper rather than unnatural and invasive. Thus this approach, as well as helping us believe in God, has ethical repercussions for our connection with animals.

In this book I argue that a plethora of new approaches to understanding evolutionary theory emphasize evolutionary constraints (Simon Conway Morris), the deep mathematics and physics that undergird evolution (Ian Stewart and Philip Ball), or the cooperative nature of much of the natural world, as well as the emergence of real novelty, as described later in this chapter.[9] A broader picture of evolutionary theory with cooperative and necessary and emergent elements, as well as the ruthless and random ones, allows us to be confident that the beauty we observe is a part of the wisdom of the creation and that the purpose that is also discernible in nature is not illusory. Although nature red in tooth and claw is not necessarily evil, there are animal traits that undergird and anticipate human violence and others that emerge in us as true altruism. Whatever is going on in evolution it is not *just* the struggle of animals in difficult terrain or competition of animals to the death followed by the victory of strength over weakness. A broader, wider evolutionary theory makes the connection between progress and predation less necessary.

Third, I suggest as part of the solution an acknowledgment that the Scriptures, especially the gospels, feature a dark world of which we speak little in the West—the world personified in other times by the names of demons or of powers and principalities, the world that gives rise to the darkest of the tares, or at least to their malignancy. This darkness may indeed be nothingness, or the absence of God, as Karl Barth suggests. But it is experienced as a power. I argue that human life is in many ways taken up into a collective and oppressive darkness, and it is beset by and cooperates with this darkness; thus the evil we experience and propagate is on a continuum with that which animals have always experienced and perhaps perpetrated. I argue that this emphasis is not only biblical but is also required to keep the distance between God and evil that theodicy requires.

Acknowledging evil as a part of the picture maintains the sense of tragedy that accompanies all life, and indeed the Genesis account, and the perception of both perfection and brokenness within the natural order long before the coming of *homo sapiens*. Moreover, the Scriptures, especially the gospels, suggest a struggle with the dark side of existence that overshadows human and creaturely life. Thus Jesus encounters Satan in the wilderness, overcomes the powers and principalities, and is victorious over death.

I argue then for an approach that combines piety and reason within a broad holistic view of nature. In the end the problem of evil will not be solved just by clever arguments, but also by our stance toward nature and toward God. Yet this prayerful and trusting stance is undermined if our reasoning is not also in line and consistent with faith.

Recent Literature

Others recently have taken on the task of reconciling evolutionary evil with a good God. Notable are those by Christopher Southgate, Celia Deane-Drummond, and Michael Murray. I will interact with Southgate's *The Groaning of Creation* in chapters 4, 5, and 6 and in other parts of the work.[10] We disagree about evil, whether or not predation is a necessity in the evolutionary process, and whether this really can be described as the best of all possible worlds. We disagree as to whether the "selving" process necessarily produces the evil associated with creatures and about the connection between this and intratrinitarian kenosis. I am immensely appreciative in other ways of the fine sense Southgate portrays of the pathos of evolutionary evil, and in every respect I agree with him also concerning both the redemption of the animal world—at least in some sense perhaps too deep for us to understand—and with his mandate as to our place in this world before redemption.

Celia Deane-Drummond has also written a very fine book, *Christ and Evolution*.[11] This work makes new strides into theologizing in the area of theology and science and in appropriating sources of wisdom already in the Eastern tradition, one that has always emphasized more the loss of innocence rather than sin in the fall and also the *Christus Victor* theme in atonement.[12] In her recent work Deane-Drummond poses the possibility of the shadow sophia from the theology of Sergii Bulgakov. Shadow sophia acknowledges also the dark side of the evolutionary process.

Michael Murray's book *Nature Red in Tooth and Claw*—with which I interact in chapters 3 and 4—is an exhaustive look at the philosophical options in theodicy, given at least the possibility of the brutality of the evolutionary process and the suffering of animals. Theologians, as I argue later, must be cognizant of these philosophical approaches, but theology brings to this task both a higher standard of defense and more resources. It is these biblical resources that I hope will give another perspective on talk of God, which is after all a theological domain. I also take more note of biological challenges to a strict neo-Darwinism than do many theologians, although recent articles and pending books by Sarah Coakley are also challenging a single focus in the evolutionary process.[13]

Behind these recent publications stand the works of John Haught, Bob Russell, Andrew Linzey, John Polkinghorne, Arthur Peacocke, Holmes Rolston III, and indeed many others.[14] For those of us who set sail in the waters of predation and animal suffering, of contemplation about fall and redemption, it is always with the work of these authors in the mind. Haught has pleaded that we look at the natural world as multilayered and mysterious rather than as mechanical and reductionistic. He has urged us always to go "deeper than Darwin." He has attempted to be true to the grammar of Genesis, suggesting that the perfection of which it speaks may be eschatological rather than historical and that evolution contains an essential openness to the future, which is the biological and theological direction of redemption.¹⁵ John Polkinghorne has long championed the ongoing search for truth in theology and science. He admits that the problem of evil is the most disturbing evidence against God, and he proffers some form of the free-creatures defense. Arthur Peacocke holds to some version of this being the best of all possible worlds. He understands evolution through the lenses of God's immanence within the evolutionary process, a process that is always in tension between chance, which is optimizing creativity, and law. Holmes Rolston III was pivotal in giving back value to the natural nonhuman world. Evolution, although Darwinian, is creative and "remembers" and learns from the past. He is and was a leader in environmental Christian ethics.

Robert Russell has had a long career at the intersection of faith and science, especially physics. He emphasizes the contingency of the natural world. He looks also to quantum mechanics as a possible locus for what he calls Non Interventionist Objective Divine Action (NIODA). Quantum mechanics is essentially open, but it also becomes possible to imagine that in this mysterious place God might influence pervasively without in any way contravening "laws" of nature. Russell, however, traces the lineage of evil in the biological and human realms down to the level of physics. This leaves him, like Haught and others, emphasizing eschatology as the answer to the problem of evil. For him the Kingdom of God and God's salvation should be understood in a radically inclusive sense as meaning changes in even the basic laws of physics, like that of the second law of thermodynamics.¹⁶

Andrew Linzey has given animals back to the theological world. He has shown us their great dignity, their humane qualities, their kinship with us, their importance to God and has single-handedly over many decades put animals back on the agenda, insisting, when this was very unpopular, that a fallenness afflicts the natural world as well as the human and social worlds. Linzey has had a prophetic voice, upbraiding also the ecotheological movement for its uncritical representation of the ecosphere as necessarily good, even when manifestly vicious.¹⁷

The Journey to Contemplate Animals and the Evolutionary Process

We begin the journey with all these thinkers before us. But first I must begin with a personal note and an apologia for this journey. Many who have known me in the past would be surprised that I am writing about animals. I am not naturally an "animal person," though I have recognized at a distance the companionship they bring. I have had to learn to journey with animals. I remember very distinctly as a child being told that the salvation story was for humans, and that the spirit of animals at death went back to the dust. Afterward I switched off from the animal world to a large extent. I was sufficiently theologically oriented that I reasoned if animals were not of ultimate importance then why worry about them now? There was already enough to concern me in my Roman Catholic world in which the process of salvation for humans appeared so fraught and so risky.

More than ten years ago I taught a class, Biology and Religion; Questions at the Interface, in which a group of liberal arts students in North Carolina—both religion majors and biology majors—met to discuss some of the issues in evolutionary theory. There was a young woman from Washington, the daughter of a pastor, who steadfastly refused to believe she was a mammal. The biology majors took her to task, quite firmly but kindly. The class became a prolonged meditation on "being an animal." I became more aware of animals, their suffering, and the theological importance of that suffering. The next decade saw leaps in our understanding of our deep genetic proximity to animals, and indeed to other forms of life, and the works of Jane Goodall, Ian Tattersall, Frans deWaal, and others brought us new glimpses of the moral and emotional life of animals. In all of this I was awakened also to the presence of animals in Scripture, a presence I had largely not heeded in the past.[8] The ecological crisis showed us our common fate, our common predicament, as the present ecological catastrophe unfolded.

Suffering, Evil, and the Existence of God

The suffering of animals for long eons and the extinction of whole species—perhaps most species—were linked with my previous worry about theodicy. The animal world came into focus at a time of great skepticism about the claims of traditional theism. Yet the orthodox systematic theology I was reading often assumed the traditional fall story somewhere behind the scenes. The core theological explanation for evil was undermined at the same time we were opening up to new vistas of evolutionary evil, but theology was not yet really grappling with this problem. The taunts of Richard Dawkins resound over this first decade of the twenty-first century, especially in the wake of 9/11 and the Boxing Day Tsunami of 2004. At those moments, prayer seemed to be lost. God turned God's back, even as we now realized we could no longer blame ourselves for all the evil that exists. Somehow or other when we could blame humans there was always the possibility we could change and do better, and humans would be in control again, or at least perhaps through prayer, in control of God. At the same time we had to face up to the cruelty of a process upon which we had long not focused.

Our remonstrance that God is there anyway can sound more like special pleading. Yes, there is no empirical evidence of God in the evolutionary process, but God is there undergirding it all, the believer says. God has purposes for creation even though none is visible in creation. We look at Scripture and insist that God is so different from the God of popular opinion, but why does this popular opinion emerge from a Bible-reading culture? When 230,000 people perish in a moment in a tsunami we become numbed to disaster and hope; nevertheless, that God is in it in ways that surpass human understanding. Sometimes it is the kenosis or self-emptying of God we blame for suffering, but then we claim God's presence when things are good. For many the list of special clauses has become too long and too incredible. The answer that says there just isn't any God that cares becomes more plausible.

I became aware that in previous centuries and indeed millennia, whenever the notion of God was threatened there was always the evidence of nature, the natural response to the beauties and harmonies of the world, that brought theologians and faithful people to admit that God must exist and then to read the Scriptures with generosity and charity, seeing what John Calvin referred to as the *divinitatis sensum*.[19] In the twentieth century and beyond, this natural sense of God has been undermined as we have been taught to be very skeptical of invoking God in any natural process. Intelligent Design has become a code word for the worst kind of underground creationist.

As a previous student and teacher of mathematics I have a particular view of the ordered patterns of the world as uncanny and disarming. Why does the square root of minus one mean anything at all in physics? Why do we find Fibonacci series in nature everywhere? Not only at the level of discernment but in this deep (mathematical) language of God I have sensed what can be understood as the shadow of love.[20] But at this level the discernment is personal and shadowy and hidden. I continue to believe because the goodness of the wheat of creation and the emergence of personality and other novelty I experience as derivative of a more perfect beauty, and oriented towards a transcendent point. This response is perhaps akin to Schleiermacher's sense of absolute dependence.[21] I am convinced, albeit at a deep and inchoate level, that personality and order and beauty are made in the image of something, that these phenomena did not emerge—with attendant laws—*de novo* 12 billion years after the Big Bang.

Conclusion

In summary, I have three aspects to my argument, all written against a backdrop of prolonged and "natural" animal suffering. First, I argue that the phenomena we see throughout nature are a mix of the perfect and the corrupted—the wheat and the tares. At times we may be able to glimpse the good, but never easily or unequivocally. Second, I argue that humans are not to blame for *all* evil because so much of it preceded human becoming; humans and perhaps some animals cooperate with evil that moves beyond and above the human level. Third, I argue that believing requires a confidence that when we look at nature we are seeing the work of God, albeit the infinitely subtle and almost infinitely hidden work of God. This work of God, reflected deep in our consciousness as *divinitatis sensum*, is what helps us say we don't understand the evil, but God's works are manifest—the response of Job. Thus this third aspect depends upon an affirmation that there are other ways of seeing the evolutionary process that are not just "nature red in tooth and claw." In conclusion, I argue that an ethical response to animals comes directly out of theological understanding of the world, especially of its history, evolution, and relationships.

In the first chapter I examine the problem of evil in general terms, given animal suffering, and its connections with the grammar of death in the Christian tradition and Christian piety. In the second chapter I look at human nature and its animal history and the repercussions for death and the Genesis story. In chapter 3 I interact with other philosophical writers in this area, and I discuss theological response to evolutionary evil in chapters 4 and 5. I address the wheat and the tares in chapter 6. In chapter 7 I examine theological sources for a theology of evolution in light of the wheat and tares and in chapter 8 probe newer aspects of evolutionary theory with the question of wheat and tares in mind. In chapter 9 I discuss the vexed problem of dualism that a wheat-and-tares world might suggest, along with the topic evil itself. I take up issues of fallenness and dominion in chapter 10. In the concluding chapter I ponder the repercussions for human interactions with animals and for our theologizing in general.

I

Animals in the Garden of Eden

> *I...would plead for a new "natural theology," in which scientific findings tell us something about God, and theological insights something about nature. The reason for a natural theology of this kind is that there is a correspondence between human intelligence and the intelligibility of the universe. We perceive and know more about the world than we need in order to survive in our earthly environment.*
>
> JÜRGEN MOLTMANN

AS HUMAN ANIMALS we are much engaged with warding off death and old age. Our attention is ever alert to threats. Even those who claim to be quite sanguine about the finitude of life are often attracted to the young and vital and otherwise show signs of sharing in the common human dread of ending. When discussing evolution with reluctant theological students the first question they raise is death, not death of animals of course, but death of humans. The discourse around death and resurrection is arguably the kernel of Christian theology and Christian piety; this is partly because the resurrection of humanity—as opposed to animals—so defines Christian belief, but partly also because death is tied to the Adamic theodicy, which still has a tight grip on Christian theology and faith.[1] The story of Genesis 3 and of Adam and Eve in the garden, the eating of the tree of the knowledge of good and evil, and the subsequent banishment and sentence of death is still there at the logical heart of much theological discourse, together with the atonement, often penal or sacrificial, as antidote to the fall.[2] For most of us who have any training in theology the Adamic discourse has become symbolic, mythic, and coded rather than literal. A large part of systematic theology and the church life that is derived from this theology in preaching and teaching nevertheless still articulates a tight story of the fall and redemption of humans as God's special creation and concern. This story not only proffers a faulty theodicy; it also assumes and perpetuates a false notion of the distinctions between animal and human.

Nor does ambiguity solve this problem. Edenic stories are often used without disambiguating their meaning. In much systematic theology there is so little concern for scientific issues that the coded meaning and the literal meaning, the mythic and logos meanings merge or are ambiguous.³ In many theological works the "before" and "after" words are still there: before the fall there was paradise. After the fall was the messy fallenness and evil we have now. In this chapter I examine this ambiguity and the way in which our new knowledge of animal kinship arbitrates at some deep level in Christian theology, breaking apart old ambiguities, broadening the possible range of salvation, but also undermining the theodicy of Adam and Eve that has been the glue that has previously held Christian theology—and to some extent Western society—together. Animals and prehuman history have unmasked a troubled worldview. This unmasking requires a different answer, a new theodicy, and ultimately a new way of interacting with the rest of life.

Eden as the Kernel of Systematic Theology

Stanley Grenz, for instance, although offering a fresh evangelical theology of relationality, begins his theology squarely in the garden:⁴

> The curtain on the biblical drama opens in the Garden of Eden. According to the story, we began our existence in seemingly perfect innocence. As the divine image-bearers, the first human pair enjoyed fellowship with God (Gen 3:8), savored transparency with each other (Gen 2:25), and experienced harmony with all creation (Gen 2:15–16, 19–20). Despite the bliss of the garden, Adam and Eve plunged humankind into sin...The idyllic community was shattered. They were banished from paradise...and the principle of death invaded their lives...Whatever it may say about the history of Adam and Eve, the Genesis narrative is *not merely* reporting their personal history. Instead, *the story is also about us*...The first failure has tainted the world and irreparably altered us, the earth's inhabitants. The pristine community...is gone for all time.⁵

Thus Grenz combines an historical and an existential interpretation of the "fall" in Genesis. What Adam and Eve *began* is continued in each one of us. In the piety of evangelical and pentecostal churches this merging and ambiguity are more common, even in churches that are not deliberately teaching "creation science." Ambiguity is very evident in a 2009 edited book on *Darwin, Creation, and the Fall* in which evangelical theologians make only very tentative steps toward facing the difficulties of an historic fall of humanity in Christian theology.[6] The ambiguity regarding the effects of the fall and its connection to the free will of "creatures" is evident even in David Hart's masterful *Doors of the Sea*, along with his lack of attention to prehuman history.[7]

John Goldingay has a similar double approach to Gen 1–3. Goldingay is typical in saying that he is reluctant to use the term "fall" at all, while affirming that fallenness aptly describes our world. He rejects the idea that there was an ontological change in human nature at the fall, but he does think that the fall refers to the "terrible" things that happened in the garden, the refusal to follow God and doing of the opposite of what God said, and that "the consequences are devastating for the entire future of humanity."[8] This devastating thing was not to fall from perfection but to fail to follow the command and calling that was humankind's unique potential. When he goes on to say that violence and death have something to do with this decision and happening he is not too far from the old interpretation. But he would emphasize that humans had a purpose and destiny, anew, as humans. Goldingay doesn't want to emphasize a paradise fall redemption in the old way, but he does insist on Genesis having an historical aspect, in which human decisions had far-reaching consequences. Nevertheless he does also say that we find in Genesis 3 "the same dynamic of temptation and disobedience as we ourselves experience"[9]—the existential interpretation. Goldingay therefore also mixes the two interpretations, and he insists on this very symbolic story having an historical edge. Although he does reject most of the more harmful and scientifically incompatible meanings the text has often carried, such as the presence of a paradise beforehand and the idea that humans and nature were all cursed as a result of human sin, he does not speak to the issue of how humans, with the particular animal backgrounds they were endowed with, could possibly have answered the call of God in a radically different way.[10]

The archetypal story of Christian theology is that humanity and all the animals lived in a state of perfection until Adam and Eve sinned by deliberate disobedience against God, taking the whole of creation with them into a state of fallenness and suffering and death. In this story humans have a privileged position that demarcates them so thoroughly from the rest of life that their actions alone can bring upon the created order all the sickness and violence we know so well. My contention is that this story is still the *strong central kernel* of Christian systematic theology and is alluded to whenever the drama of creation fall and redemption is mentioned. While few theologians would now take it literally if pressed, the shadow of an older meaning remains. Genesis discourse is not easily disambiguated. After all, theology students who anguish about Adamic theodicy and any threat to its integrity have studied Genesis and have learned that it was written, or at least assembled, from ancient sources either post exile or post exodus. They have heard that it may have been written down in its present form as a counter to or engagement with the Babylonians or the Egyptians or both. They know all this, but it doesn't make any difference to the grammar of Adamic theodicy. The story is still the story, and retains power and a kernel of historicity in its explanation of all suffering and death. Moreover, theology in the twentieth century did not always demand a rapprochement with science. Certainly Barthian theology actively discouraged it, and Barthian theology in this spirit is alive and well today.

Barth and Eden

Karl Barth is interesting for the way in which theology under his influence has studiously avoided engagement with science and has avoided any incorporation of the language of science into theology—brave attempts by some including Thomas Torrance and Neil Messer to rescue Barth for science notwithstanding.[11] Barth protects theology from the troublesome input of scientific knowledge. For Barth the human person is the person addressed by the Word and under the power of the Word. The human is constituted as he or she hears and engages the Word. To be a human is to be summoned by the Word of God.[12] He says, further:

> The self-understanding in which the man of sin knows himself as such and therefore as fallen man can only be the understanding of the man who hears and believes the Word of God, of the man who learns and accepts God's judgment concerning him and is ready to see and understand himself in the light of this judgment.[13]

Not only do these passages discourage theological dialogue with science, they implicitly define a relationship with God as requiring language.

Defining humanity only in terms of the upward relationship also means that the tricky questions of origins and of animal nature are not properly engaged.

Barth speaks of saga as taking up where history leaves off, of the edges where history cannot go.[14] This is partly true, that creation itself is an edge over which we cannot see, but science can and does ask questions about human origins and uniqueness. Barth's coyness renders these questions functionally meaningless. In fact Barth explicitly warns us not to go there, not to delve into the mysteries of scientific history and its relationship to sacred history:

> We miss the unprecedented and incomparable thing which the Genesis passages tell us of the coming into being and existence of Adam if we try to read and understand it as history, relating it either favourably or unfavourably to scientific paleontology, or what we now know with some historical certainty concerning the oldest and most primitive forms of human life.[15]

Barth sums up well what some writers have referred to as the independent mode of relating science and faith.[16] Each will have its own area, its own discourse, and never the two shall meet without in some sense undermining the other. So yes, Genesis should not be read as history. All is well and good. But for Barth Genesis should not be compared in any sense at all to paleontology either. One senses that Barth is uneasy with even a transversal dialogue, with any input at all from science.[17] What Barth says about the vestiges of the Trinity reveals his reluctance to engage with the natural world at all; he cautions that any signs of these vestiges in nature might create another root by which God is known, one that isn't at all based in Scripture.[18] Barth is rightly cautious at one level because too swift an integration of science and theology can lead to a reduction and simplification of both disciplines, but theology done without scientific input becomes increasingly meaningless.

More often than not, then, the central grammar of the Adamic fall has shifted into a mode whereby questions of origins are not raised at all and the existential universalized meaning of falling into sin replaces the old historic meaning. Theology in this way is removed further from the area of integration and discourse with science, and the profound repercussions of evolutionary theory and the prehuman history of the animals does not touch our theology at all.

Interestingly, though, the same belief that Christ is the center of history, and the way of truth and life, can lead in quite another direction from that of Barth. Hendrikus Berkhof, for instance, says:

> If in Jesus Christ the Father is revealed who is the creator of heaven and earth, then in Christ also the ultimate mystery of created reality must be manifest... But then it is impossible that this reality, which inalienably remains God's creation, should not in all kinds of ways, no matter how fragmentary, incidental, or broken, *bear witness to the purposes of its creator*... Then it must also be a fact that this man [created for relationship with God] in this created reality *must regularly detect signs, rays and disclosures of God's nature and purpose.*[19]

In this work I follow Berkhof—and much reformed theology and philosophy—and indeed Moltmann, with his plea for a new natural theology,[20] assuming that the deep connections evident between Christ and the matter of life will bring us back to nature and to an expectation that the creation will bear witness to the purposes of God if we know how to look. In particular the life and suffering of animals is of immense importance to our theology.

Genesis and Eden

What Genesis does mean in terms of origins and fall has been long debated. One can become quite giddy contemplating the range of scholarly opinion. Genesis is generally accepted to portray a high view of God the Creator, as one who stands outside and above the created order—though the rest of the narrative will also soften this transcendence and give evidence of God's proximity to nature. God's breath enlivens all life. God creates all life and is pleased with this life, and places humans within and above animal life in order to protect it and to image God uniquely. It is in Genesis 3 that the narrative takes a turn, and the meaning becomes opaque, or alternatively, deeply ambiguous. The traditional view of Genesis 3 is that it describes the one all-important rupture in human history, the fall from the state of grace to the state of sin, a tragedy of cosmic importance. This is the meaning that I would claim still invades systematic theology. At the other end of the spectrum of interpreters are those who would argue that Genesis has little at all to say about origins or death, or about the fall.[21] Walter Brueggemann, for instance, argues that the literalists have Genesis completely wrong, but so do those rationalists who claim that Genesis is timeless. Genesis is, he thinks, about the scandal of a God who transforms all things by the Word. He disagrees, however, that Genesis 3 has the pessimistic view of human nature that a fall presupposes. The serpent does not stand for evil and the Bible gives no explanation for evil. These statements are made with authority but without much argument.[22] For Brueggemann, Genesis speaks to the existential condition of the human being as he or she moves between the poles of worldly resistance and God's grace.[23] These interpretations, true as they must be at one level—and the profundity of Genesis is evident in its almost infinite levels of interpretation and meaning—also do nothing to encourage a dialogue with science, for they devalue the empirical and the historical; existential interpretations do not begin to touch the questions of morality and meaning that science itself is raising because the existential assumes human sentience as the locus of salvation drama. Other questions are side-lined: why do animals suffer? Do dogs know they have done wrong? Why do chimpanzees become aggressive? What level of sentience does the charge of morality require? Existential interpretations do not answer these questions for beings who have intermediate and indeterminate consciousness. Thus we are alienated from questions about our origins.

Other biblical scholars, however, do not rule Genesis out as an etiological tale. Paul Ricoeur argues that the "Adamic myth" is etiological in locating the origins of evil in the human race, but also in separating evil from the origins of good.[24] For him, however, the myth embraces the existential as well. He insists that there is no historical reference of this myth, and that the theology of fall and original sin has done untold harm in the world.[25]

Where should one stand on this continuum of interpretation? It has long been noted that the Hebrew Bible hardly ever returns to the Genesis story of origins. On the surface it appears not to have the central coding function for all existence that it has had in the Christian canon because of Paul and Romans 5.[26] Yet even on that note André LaCocque would differ from the mainstream, arguing the importance of Genesis for Jews:

> ...Genesis 2–3 has become along the many centuries of its existence as a literary composition one of those primordial warrants constitutive of human existence—at least in the world influenced by *Jewish, Christian and Islamic worldviews*. It has informed our understanding of the creation of humanity, of its proximity with the divine and its superiority over animals and plants, of the fundamental equality of all humans, of their knowledge (for better or for worse) of good and

evil, of their primal transgression and inclination to transgress the divine commandment, of their striving towards immortality, and more.[27]

Henri Blocher also dissents, saying, "the Old Testament certainly contains echoes of the Eden story audible to a sensitive ear." He hears these echoes in Hos 6:7, Ps 82, Ezek 28, and Job 31:33.[28] He continues:

> Even if the Bible contained no echoes at all of that passage, frequency of occurrence could not be the sole measure of importance; its place in the canon is significant. It is obvious that the Eden story is no peripheral anecdote or marginal addition; it belongs decisively to the structure of Genesis and to that of the Torah.[29]

Thus for these interpreters Genesis is not easily explained away. Its enigmatic and often misunderstood presence hovers over the Christian story and our theological interpretations of existence and of the orderings of life. There remains a strong psychological presumption, backed by interpretations such as those given by Blocher and Goldingay, that Genesis is relevant in some way to our discussions on origins, that the mysteries of life and death, of what it means to be an animal or to be human are encoded deep in its prose, and that the scientific and theological stories must be brought into some form of dialogue.[30]

I would agree, but not because I think Genesis 1–3 is historical in our sense of the word. It does not need to be historical in any way to impinge on matters of history. The evolutionary story, rooted as it is in empirical observation and our experience of animals, nevertheless has also become mythical. The resonances of these two myths must be placed in some sort of dialogue. We must sit and ponder them both and admit the deep discordances their grammars evoke. Where one gives us images of perfection, human superiority, and a brokenness brought about by human initiative, the other presents humans as one particular endpoint of a beautiful but troubling history; in this history humans are very much animals and exacerbate but are not the beginning of violence, aggression, envy, and death. Although Genesis is deep enough to absorb the shift in focus that evolution requires must be made, the repercussions for our anthropology and soteriology are immense. Thus I am arguing that the historical must in some sense shift and impact the mythical, at least in so far as the resonances of that myth are dealt with in our theologies today. Removing the discourse surrounding the myth to the existential realm alone belies the deep changes in our theology that must be faced when we begin to think in an evolutionary perspective.

Eastern Orthodoxy

Another family of interpretations of Genesis is that of Eastern Orthodoxy, which has long read Genesis 3 as a story of a fall from innocence rather than a fall into guilt—for both the individual and the race. Western systematic theologians have been interacting more frequently with Eastern theologians and with their Cappadocian sources.[31] A fall out of innocence has some coherence in an evolutionary paradigm, though it would appear that *homo sapiens* was probably not exactly innocent at the threshold of the species. Nevertheless there are hunter-gatherer groups today who would appear to be much more innocent than humans in the developed world.[32] There may indeed be, as LaCocque implies, an inverse relationship between our scientific knowledge and our innocence.[33]

There have always been interpretations also that understand Genesis 3 more as necessity than tragedy, more a coming of age than a fall from grace. The element of a necessary period of testing or trial is one that appears to make sense of the human experience, but it does not do justice to the tragic dimension evident in Genesis and indeed throughout Scripture.

One readily recognizes that biblical scholars are as divided as theologians on these issues. I have chosen to interact with those scholars who have given new ecological readings of Genesis, or with those Hebrew/Christian scholars who write with an understanding of the predicament older interpretations have brought to theology in dialogue with science, and who can give new understandings of evil. We can all agree that Genesis has had a powerful and world constructing role in Western society and church; it has deeply shaped our consciousness and our sense of self in the world, and it has accentuated in most cases the gulf we feel with other species. Genesis has also profoundly formed our responses to evil and to the darker aspects of the natural world itself.

Animals, Paradise, and Society

I shall return to the vexed problem of Genesis in the next chapter and in Chapter 10, where fallenness is discussed. In the meantime, suffice to say that it is animals that will force an interpretation on many issues one way or the other. Animals have made us question Genesis-derived assumptions about animal inferiority, animal morality or its lack, and animal reactions to death. Our consciousness of animals will also compel us to read the biblical text differently and to read theology with this new understanding of animals in focus, a change that has already begun in the new ecotheological consciousness. In recent years we have begun to know animals better, not just by mapping their DNA and learning their objective characteristics, important though those are, but by seeing them more personally as creatures with their own sense of being alive and their own emotional range.

Aside from theological considerations, what blows apart the old tight story of creation, fall, and redemption is the close observation of nonhuman animals. The longer we look at our animal cousins, at their DNA, their morphology, their behavior, and their social practices, or at primate culture, the more we see ourselves and certainly the connections with our predecessors. As J. Wentzel Van Huyssteen has said, in attending to animals we become convinced not only of the connections between ourselves and them, but also of our uniqueness, but the uniqueness is a qualified function of a deep kinship.[34] This conviction, of our connection to animals *and* our uniqueness, has been lacking in so much theology, though this misunderstanding in many ways reflects a general social ethos of confusion around the distinctions and continuities between the animal world and the human. If, however, morality begins and is laid upon a substrate in the animal world, if sinfulness at least in latent form begins with animals, then the traditional story of creation, fall, and redemption can no longer hold.

The most important sense in which we fail to attend to our animal history is in ignoring or bracketing human evolutionary connection to animals. Christians are not alone in this. Since Darwin we have failed to live within or construct in church or society, a coherent story of origins that makes sense of morality. The paradoxical combination of distaste for and fascination with our hominid origins is widespread. There are, for instance, two major views of human behavior—the social science model, which sees behavior as liberated from our evolutionary past and almost entirely socially constructed, and the evolutionary model, which reduces all human impulses to their evolutionary survival value. In spite of recent science a tenacious association lingers between beasts and violent or subhuman or sinful behavior. LaCocque points out that aggressing nations in the Bible are often referred to as vicious beasts, referring to Ps 22:13, 16,

20, 21, Daniel 7, and so on.[35] Today one hears talk of animallike behavior often in a court of law, applied particularly to horrendous crime.

Thus the confusion about our connection to and difference from animals is widespread and extends beyond the Christian worldview. Only recently have we begun to study wild animals in their own habitats and to realize how false our preconceptions are. On the one hand, we now know that most animals kill only to eat, even if the killing is vicious and sometimes prolonged. Higher animals also care well for their young, with notable exceptions. Although there has been a widespread connection between bestiality and human maleficence, most animals are not bestial at all. On the other hand, aggression is not absent from the animal domain. Some of the higher primates are somewhat aggressive. From primate studies we are told that whatever violence we have did not emerge for the first time with our humanity, but before. In some of the higher primates there is infanticide, murder, aggression, and occasional warfare.[36] This behavior, however, is so recently discovered that it cannot account for the reputation of animals in the human world. On the whole, although we can see some of our traits reflected in them, they are not anywhere near as violent as we are.

A Christian critic might say, though, that perhaps all this animal violence is an indirect result of human ravaging of the ecosphere, that animals in this way have also been affected by the "fall."[37] There are examples of this connection. At the very least, however, anthropogenic effects on animals show that animals can be affected physiologically in very similar ways to humans. Moreover, we know that predation has long preceded human becoming and it is the skills learned in predation that are used in aggression. Another point here is that animals show not only violence and aggression but also compassion and care. Neither extreme was at one stage thought to be truly present in the less-than-human. If compassion is there it will not have been a recent emerging, and so we might conclude that the same would apply to the aggression.

And indeed Christianity has been most reluctant to ponder our connection with animals, especially in Protestant modernity. The story of salvation begins with Abraham, continues in the chosen people of Israel and their dramatic salvation through the Red Sea, and climaxes in salvation of humanity in the crucifixion and resurrection of Christ. Although Jesus is known as the Lamb of God, the drama of salvation has rarely considered animal life or indeed the rest of the universe. Only recently has that changed, especially with the development of animal theology and particularly ecotheology. Andrew Linzey, the foremost proponent of a theology that is animal centered, argues that it is our theology of God that influences how we understand the rest of creation. A theology of God that allows that God suffers with the creation is more likely to take the nonhuman creation into account. He says, "if it is true that God is the Creator and sustainer of the whole world of life, then it is inconceivable that God is not also a co-sufferer in the world of non-human creatures as well."[38] It is also our theology of God that leaves us easy or uneasy with the idea of God-intended carnage and predation throughout the evolutionary process.

Human and Animal DNA

The scientific evidence for human embeddedness in evolutionary history has climaxed in the last decade or so with the mapping of human and primate DNA. This information, which gives us incontrovertible proof of our origins, is found in comparisons of the chimp and human genomes. Clearly some scientists were expecting a much, much larger human genome and one more radically different from other animals.[39] Surprisingly, our DNA is, as is well known, 95–98 percent the same as chimpanzee's. We now know that small changes in DNA can have large effects on the organism or even result in speciation.[40] Nevertheless, the "posthuman genome" view of the world is of DNA that is largely common to all similar species. No wide gap of information separates us from our primate cousins, or even from more primitive forms of life.[41]

DNA is not everything by any means, as I will discuss in Chapter 8, but we are increasingly aware that the same simple formulae are used to make us that also make other mammals and primates. There have been not so much additions to our DNA as reconfiguring of previous genes, with the slow addition not so much of new information as insertions of jumping genes and viruses, fusions and translations of genetic material. The HAR1 gene, for instance, is one of the genes that has undergone selective pressure since the time of the split from the chimp line. Yet it was always a part of the genetic landscape.[42] Graeme Finlay has given us a fascinating glimpse of genetic similarities among great apes, including ourselves, and of the way evolution works at this level.[43] In *Homo Divinus* and other articles Finlay has plotted the insertions of errors, fragments, viruses, and jumping genes into the family tree of primates and mammals at different points. In a method similar to that of biblical textual criticism, these insertions can be used to plot the lineage of humans from great apes and down even further into the mammalian line.

This has given us a great deal to think about. One interpretation is that evolution has acted a bit like a multidimensional Rubik's cube, where every arrangement is meaningful but where the final permutation is the one we now have of modern *homo sapiens*. Thus evolution prehends or anticipates what is to come.[44] Klapwijk describes anticipation in this way: "Thus there are good reasons to state that developments in living nature too represent an anticipatory process, an anticipatory process in the sense of a directional or purposeful development but without the implication that the purpose is known."[45] Humans are connected with all other life and anticipated by it. Life is a more interconnected process than we might once have imagined.[46]

Conclusion

In this chapter I argue that the tight connection between death, morality, and Eden has been unraveled as the manifold connections and continuities between animals and humans are revealed. Much systematic theology continues without much backward glance at the state of science. This is true of the traditional paradise/fall/redemption grammar but also of existentialist theologies that ignore science altogether. This area is confusing because Genesis itself is interpreted in such a wide variety of ways. Animal science is now making a difference, however, and should make a difference to our theology. Animals not only die, and have died for millennia before humanity, but they are corrupted in part in the same way that humans are, with tendencies toward aggression and rivalry and certainly selfishness. The sharp theological exclusive boundary between humanity and animals has been broken. The emerging story of animals unravels old theodicies and old explanations for the tragic and for death and suffering. In the next chapter I reexamine the moral and affective links between humanity and animals and the ways in which, perhaps, animals share in the sting of death. The Genesis story will be revisited and reread against the backdrop of our close biological connection to animals. And I raise the question of what evolutionary history, especially the history of sentient creatures, does to our notions of theology, God, and evil.

2

Humans, Animals, and Death Revisited

> *Being both more systematically brutal than chimps and
> more empathic than bonobos, we are by far the most
> bipolar ape.*
>
> FRANS DE WAAL

IN THE PREVIOUS chapter I pondered biological and evolutionary evidence for our close kinship with animals and the way in which the old grammar of Eden, death and fall and associated theodicy, are unraveled by these links. Animals emerge as more like us than we might once have suspected. In this chapter I look at human and animal nature and the evidence for proto-moral life in many of the higher animals. I ask whether the sting of death is not present also for animals to some extent. I return to Genesis 3 to question how we might interpret it in light of our new knowledge. Death and suffering emerge as outstanding realities for both human and animal; our theological certainties of the past and solutions to these problems, however, are in disarray. This chapter therefore opens up the question of how to answer the problem of evil given what we now know of animal suffering and our own animal history.

Animals and Morality

In the last century or so there has been an ongoing and persistent search for human origins and for the mystery of what it is that makes us different from and similar to the animals. The modern search begins with Darwin, who saw the parallels in emotional life between animals and human species.[1]

We now know that random mutations and aggressive instincts are not all that we share with our animal cousins. A whole range of axial emotions like empathy and reciprocity and mutual enjoyment, which are the ground upon which higher moral emotions flourish, is present in chimps, baboons, bonobos, and many other animals. Similarly, although no other animals appear to have anywhere near the complex symbolic language of humans, most animals communicate; higher primates communicate in rich and complex ways with grunts, signals, and intricate bodily expressions. The extraordinary complexity of our language must once have built upon such an underlay.

Animal studies now show us that we are not anthropomorphizing when we ascribe emotions and indeed empathy to animals. Frans de Waal is one of many primatologists who have studied wild and captive populations of chimpanzees and bonobos, discovering that similarities are more evident than they once were. De Waal also studies these animals with an eye to how they can be interpreted in terms of human categories of morality and social organization. De Waal argues: "Being both more systematically brutal than chimps and more empathic than bonobos, we are by far the most bipolar ape."[3] He also names us the "janus" ape, because "we are the product of opposing forces, such as the need to think of our own interests and the need to get along."[4] De Waal notes that we have inherited both aggressive and deeply cooperative tendencies from our ape precursors. In chimps, however, the aggression is kept more often in check by elaborate rituals of grooming and making up and by close proximity. In humans these same tendencies become malicious with increasing human social isolation and under the escalation that ideology and technology have made possible in human societies.

De Waal observes the "golden rule"—do unto others as you would have them do to you—operating in primate colonies and in communities of other animals such as dolphins and elephants because these animals are able to put themselves in the position of another. The golden rule is something we have also made more conscious and religious, enshrined in the moral codes of most religions. Humans have more capacity to decide for themselves whether or not they keep the golden rule. Yet its application is widespread at various levels in the animal world; not only humans but primate communities also help those among them who are weak or crippled or intellectually handicapped.[5] Moreover, even the most hierarchical chimp community is flatter—if more conspicuously structured—than our contemporary societies are.[6]

This animal heritage highlights the way in which human moral capacities have grown out of these preliterate capacities and are endemic in so-called lower animals. Humans did not invent the social dimension; this is a part of our biological inheritance; the social comes with the bodily. Theologically, of course, this bodily social capacity should make the idea of a Trinitarian God less arbitrary and abstract. How odd it would have been of this God to make in God's image essentially lone individuals—although rare instances of such exist in the animal kingdom (the rhinoceros and the coyote, for instance). Moreover, although we are instructed to love as Christ loves us, and Christ made love perfect, lower forms of love were never alien to life, but rather were embedded in the evolutionary progression.[7] Even in biology these findings are relevant. The idea of the "selfish gene" and of nature *only* "red in tooth and claw" becomes a misleading simplification of the reality of cooperative life forms at all levels from that of the amoeba to the human. Humans are more tempted to self-interest than are other animals, but the capacity and the predilection for altruism is a part of primate and mammalian life.

The continuities do not eliminate the genuinely new that human and indeed each animal level constitutes. The danger of pondering our animal inheritance can be a tendency to reduce all behavior, both animal and human, to the end result of mechanical evolutionary outcomes. Humans are not *just* more developed animals, and animals are not *just* on the way to higher sentience. Ian Tattersall, for instance, summing up the now voluminous literature on this subject says: "The cognitive gap between them and us has been narrowed somewhat by [such] studies...but it is far from closed—and obviously never will be."[8] There are always two ways of looking at animal observations, marveling, on the one hand, at how much is already present in primates and, on the other hand, acknowledging that we have something unique in symbolic language and associated powers, together with the rich overlay of cultural and political law and custom that no other animal yet has.[9] Similar ambiguity surrounds the sense of self that appears to be one of the apparent preconditions of a particularly human self-consciousness. The differences that characterize our remarkable brains occur in subtle arrangements and reorganizations of the architecture of the brain. Although we are "different" the reminders of where we have come from are present in our brains, our chemistry, our culture, and our language.

Tattersall is concerned with the process of "becoming human," but along the way he also cautions that, shared traits notwithstanding, living primates have their own way of being in the world that is distinct, and this is not just a lesser form of humanity. To see all animals as climbing a ladder to human status is wrong, he claims.[10] Nevertheless it is also wrong not to acknowledge the ongoing presence of our earlier brains—and associated behaviors—within the modern human brain we have today. In the past, and even very often today, this is the error religious people are most inclined to make, believing there is an absolute divide between us and other animals.[11] Thus human greatness, human creativity, and human moral strength are all shown to grow out of a rich tapestry of traits and emotions that we share with our animal cousins.

Animal Self-Consciousness

In some ways all of this research raises as many problems as it solves. But we can deduce that the boundary between humans and primates is more permeable, less well defined than was once the case. Animals, for instance, can be separated into those who have self-recognition in a mirror and those that don't, because this appears to be a test for a highly developed sense of self. The instinctual caring evident even in lower animals becomes a more consciously intended caring and altruism with higher levels of self-awareness. When an animal can recognize itself in a mirror it also appears to have a highly developed sense of itself as different from other minds, and the first glimmerings of an understanding of how minds other than themselves work. At least this test, which divides humans, chimps, dolphins, and elephants from other animals, including dogs and monkeys, is one of the best we have yet of this dawning sense of self-awareness and the beginnings of a moral life.[12]

Empathy isn't just an ill-defined feminine concept; it has a neurological basis in mirror neurons, common to all mammals.[13] The discovery of mirror neurons gives another objective backing to the thesis that humans and animals share a great deal, and should finally put to death the idea that animals don't really feel emotions. Mirror neurons make sense of the empathy we feel for animals, and they for us. They also make sense of the obvious web of caring and concern that surrounds an animal community much as it does a human one.[14] Along with empathy, however, comes also the dark side of human sentience, present to some extent in chimps—mimetic desire and deliberate unprovoked violence and aggression.

Thus the highest of human emotions is not only shared by a few lower animals but is tied strongly to recognition of the bodily form as self.

Where once empathy may have been seen as that which enabled the flight of the human from the bondage of the body, recent science suggests the reverse is the case; empathy is rooted in deep evolutionary mechanisms by which life becomes possible in social groups. Empathy is that which makes all moral life function, all relationships and everything which truly differentiates us from clever rule-bound machines. Nevertheless there are limits to empathy. Some animals have been documented coming to the aid of the young of other species, but empathy tends to break down at the borders of tribes, as indeed it does with humans.[15] Slight differences in bodily makeup signal "otherness" and can result in predation or aggression, especially in chimp colonies, where other monkeys are often used as food in violent ways.[16] Even between chimp colonies the borders of the tribe's territory are patrolled vigorously, though this is not so much the case for bonobos, a smaller form of chimp and also a cousin of humans. We should not find it surprising that humans are also hardwired in similar directions. Overcoming these body-linked prejudices does not come naturally; rather it requires recognition of their deep evolutionary roots and their links to subtle differences in bodily make-up. In order to overcome prejudice we must think critically about the capacities we have inherited. Unless humanity is very different from animals closely akin to us today, we did not come to the boundary of the species innocent. The "wheat and the tares," the good and the evil, were there in some form from the beginning.

The Neo-Cartesian Promoters

Nevertheless, not everyone would agree that we are tied to animals by links of empathy and shared emotions. There is still a contingent of philosophers who argue that animals do not experience emotions in the same ways that we do and who then conclude that animals essentially do not suffer as humans do. Death and suffering in the animal world, they conclude, have no moral significance. Michael Murray sums up these arguments in chapter 2 of his book, *Nature Red in Tooth and Claw*. René Descartes is well-known for his conclusion that animals don't really suffer, but instead have mechanical processes that make them appear to us to suffer. Murray discusses the present day neo-Cartesians and concludes that they do have a strong logical argument but he suspects "...that few will find the neo-Cartesian position to be compelling or even believable."[17] He suspects the natural anthropomorphizing tendency of humans is too ingrained to be thoroughly dismissed. I would argue that there has tended to be a complex mix of anthropomorphizing of animals, on the one hand, and neglecting them, on the other, but that the research currently is swinging in favor of those who see links in mindfulness, sentience, and morality between the species.

Theology can have some significant input here, because although as people of faith we have neglected animals as part of our story of salvation, when we start looking they are all over the sacred Book. Surely we would not be told that restoration will be like it is for calves released from the stall (Mal 4:2) or that the Lamb of God would go to his death like a "lamb to the slaughter" (Isa 53:7) if animals were only place-holders on the way to human salvation. Surely the metaphors for Jesus as "being led as a lamb to the slaughter" and the "Lamb of God" would not be a suitable if animals did not share at some level in the emotions and proto-morality that humans enjoy. Nor would there have been a covenant made between God and all flesh (Gen 9: 10, 12, 15, 15, 17) as John Olley has pointed out, if all flesh were not important to God.[18]

Moreover, we often experience from animals some of the same clues as to the inner life that we do in humans—signs from dogs, for instance, that they feel remorse. In humans we have a backup in speech but we rely very often on these nonverbal signs, even to the extent that we do not believe a person is really repentant unless they exhibit some of the nonverbal clues as well. In the case of humans we believe that these nonverbal signs are less susceptible to deceiving manipulation than are words alone. Why then would we not heed these signs when they do come from animals?

Blank Slate View of Humanity

None of this is widely appropriated, certainly not in the church. Although humans have always attributed emotions to domestic animals, and the odd lively and intelligent wild variety, we have been willing to treat animals in a way that was more consistent with a belief that they were totally "other." A result of this lack of attending to animals is that we often perceive our human nature to be entirely ours alone, and with the accompanying view that we are born as blank slates. Mary Midgley argues that this studied neglect of animals and our connection to them is because we have not internalized evolution. She also sees it as a reaction to modernism. In a bleak modernist worldscape humans are the only ones who lived outside that.[19] To claim too close an association with the natural would be to threaten human freedom and autonomy, and perhaps even human responsibility, she argues.

Existentialist views of human nature add to this slate the accumulated effects of our own free actions. Postmodern anti-essentialism also insists that the self is constructed and hence denies the importance of biological hard-wiring. This cluster of views of humanity rejects the idea that humans have a given, inherited human nature or set of dispositions and preferences. Theological variants of this position assume a sinful nature for humanity, but only because humans are fallen. All of these are ways by which the freedom and difference of the human is asserted and the givenness of human nature is denied. Thus the animal aspect of human nature is denied, and especially an animal inheritance that is already balanced toward the aggressive, toward mimetic desire, status seeking, and selfishness, as well as empathy and altruism.

Where this worldview has been challenged is in the reductive theories of evolutionary psychology in which human traits are explained in terms of their evolutionary advantage alone. But Nicholas Healy has also argued that in this postmodern technological world we now have a belief that even the human genome and the biological basis of our inheritance are infinitely malleable and changeable—that we are now in charge of evolution. He believes that this gives us a sense of being trapped inside a mechanical system, or alternatively that our freedom places us outside the jurisdiction of God.[20] Both believers and sceptics are caught up in these worldviews. Yet increasing numbers of biologists and theologians are arguing that we are animals, and that we cannot afford to deny the links. Similarly, though, animals are more like us than once we thought. Humans are not blank slates at birth (even if we are free in ideal circumstances to be endlessly creative and adaptable). Apart from all other considerations, Mary Midgley says, to think this way is to proceed "as if the world contained only dead matter (things) on the one hand and fully rational, educated, adult human beings on the other—as if there were no other life-forms."[21] Midgley argues that this flat worldview has proved fatal to other life forms, especially to animals.

There is of course, some truth in existentialist and social views of human behavior. There is a moment of choice, and there is a threshold in each life as one passes from a state of relative innocence to one of conscious choice. But the choosing self is a given, is inherited, is already a set of emotional and physical preferences, and these extend backward deep into our primate beginnings. There is also truth in the fact that human nature is a malleable and creative and outward focused, and that this may be what makes humans human. What Van Huyssteen calls the fluidity of our minds and what Pannenberg refers to as exocentricity, or openness to the world, is characteristic of us as humans but so also are the predispositions and tendencies we have derived from our animal ancestors.[22] Reasonable also is the postmodern insistence on the human capacity for social constructions of reality. Perhaps alone among the animals we can project meaning upon the universe; we have become so enthralled at this capacity, however, that we think this make-believe is all there is.

These definitions of fluidity and exocentricity point to continuity and discontinuity in humanity's animal inheritance. All of this is compatible with the givenness of our animal inheritance. It is this nature we have inherited that has acquired fluidity and malleability. Meaningful gesturing and residual language in animals have been adapted by both mathematics, which often includes a more realist perspective on the world, and anti-essentialist postmodern deconstruction. Thus to say that we have a given, human nature is not to say that we are determined, nor that we must behave a certain way; rather it refers to a balance, a predilection, a vulnerability. "It means," says Midgley, that humans are "aggressive among other things."[23] The beginnings of the ingredients of human rationality and capacity for violence and creativity go deep into humanity's animal cousins and primate ancestors. This is a part of the human story and part of theirs. Emphasizing our animal inheritance can leave us, as humans, convinced that we have no freedom and live only within biological constraints; if we assume, however, an emergent evolutionary process, as Clayton and Klapwijk and others have described, then the uniqueness of human freedom can be seen as an emergent and very real characteristic without denying evolutionary contributions to our human nature.

Nevertheless, there are no easy structural boundaries between the ape precursor and the hominid who was to believe it was made in the image of God. Species differentiation may take a very long time in human terms.[24] The innate abilities that culminate in the kind of rationality and spirituality humans possess long predated us in some kind of primitive form. The shadow of the *imago Dei* stretches way back. Of the twenty or so hominids around on earth in the 2 million years preceding us we are the only ones left.[25] Modern humans and *neanderthal*, another sapient species, lived side by side and are now thought to have interbred to some extent in Europe.[26] Whatever was coming to being in us was represented in a branching hominid tree, of which many branches have been dead ends but could potentially have developed further. The questions inherent in the two poles of our understanding, our special status under God and our evolutionary contingency, must be held in tension.

Only now is the depth of our animal past explicitly apparent. Attending to this story makes us take more seriously the eschatological images of the lion and the wolf lying down together (Isa 11:6), and of the whole cosmos groaning as its waits in anticipation of its redemption (Rom 8).[27] Affirming and accepting this animal inheritance and the way in which significant ingredients of what makes us human were already in place in the species that preceded us does change and challenge some of theology's key doctrines of salvation, fall, and *imago Dei*, and hence challenges our theodicy.

Rethinking Death in Light of Animals

Thus the close connections that are now evident between human and animals do make clearer the path ahead in theology and theodicy. Humans are not the original or at least only causes of death, nor of violence, selfishness, aggression, and anxiety; this may make us more attentive to the outside source, as André LaCocque calls it, even a prior source to evil. "Revolt is not limited to the human realm, but is a universal phenomenon...," he says.[28] How does this mix of continuity and discontinuity in emotions, morals, aggression, and altruism and relationships to mortality change our understanding of death? We used to understand death; humans caused death by their rebellion and Jesus came to save humans from this very fate. This only made any sense, of course, within a very narrow time frame. Now we must grapple with evolutionary time scales. Now we must ask whether animals too share in the sting of death, whether they too therefore suffer and whether they too are a part of salvation? If they do suffer we must add that suffering to the concerns of the theodicy problem and our questions about the nature of God.

First, it may be the case that animals share to some extent in what we call the "sting of death." If the sting of death to which Scripture refers is related to the complex interaction of sin and death, both physical and spiritual, resulting in a dread and fear of death, animals must share in this to a degree. Indeed Gen 9:2 refers to the "fear and dread" that all animals will experience toward humans in a postdiluvian world. Humans have long assumed, however, that these emotions were not properly a part of the animal repertoire. Recent reports of the behavior of the female companions of an elderly female chimp who was dying are therefore interesting. They show increased attentiveness, vigilance, grooming, and testing for signs of life. The daughter of the dead chimp slept the night by her side. The chimps all were subdued and sad following the death.[29] Animals, too, live lives that miss the mark of perfection, one of the major biblical definitions of sin;[30] they too are entangled in the law of the jungle, and react to danger with a similar fear to that of humans. For humans there is a particular awareness of death and fear of death—seeing very acutely the existential edge that it demarcates, having no itinerary for postmortem existence, no real reports from the dead apart from the resurrection and that mysterious, it being essentially a journey away from home and all that this conjures up for the human. All of this is a part of the distinctly human aspect of death, though the chimpanzees in these recent reports obviously come very close to the human experience; that elderly chimp certainly knew a companionship and care many humans do not in old age. For the contemporary person the journey to death is also taken in the midst of a reality that is flatter than has previously been the case. We don't see the depth of nature, or its multiple dimensions and its hints of "otherwise."[31]

There may indeed be differences in this regard between human groups, with increasing "sting" following increasing sophistication and technology. There does seem to be a particular burden of death that we share as modern Western people, perhaps evidenced in the relentless movement of armies and colonial domination and in the flat mechanical imprint we place on all we see and discuss.[32] Yet if there is a range of human reaction to death there must also be some aspects of the sting that animals share. Animals do not make a home out of the fragility of culture and artifact, but they tenaciously hold onto life; all animal life forms do so. Animals know when they are cornered; they know that predators threaten their very existence. They also show physiological signs of fear and of increased stress. Thus long before humans and the non-human/human boundary and the possible "eating" of the tree of the knowledge of good and evil animals, we must assume, suffered and fought for life and died; they grieved, as elephants and chimps do today. Whatever death in its physical and even emotional sense is, it is not connected only to the fall of humanity. Animals have suffered and died and grieved as a part of the regular fate of all animal sentience long before humans walked the earth. This death that so captures the imaginations of many Christians today is indeed an integral part of the living process as we know it. For many animals it is regretted and resisted; we must consider at least untimely or diseased or suffering animal death evil in the way we consider much human suffering and untimely death also to be evil, its being "natural" notwithstanding.³³ The complexity of death in the animal world enormously complexifies the problem of evil and the Christian story. These aspects of animal death also help us to adjudicate between possible interpretations of Genesis and ironically open us up to more resonance with animals in Genesis, as I discuss in the next section.

Second, what then of spiritual death? Is this the provenance of humans alone? Humans have over animals a highly intense awareness of the meaning of mortality, together with a sense of the possibility and perhaps even desirability of immortality. Humans make a home that must then be given up. Animals live more easily within the confines of the natural world. Animals also need little training to engage the complexities of social life in the tribe, something that humans must forge by painful efforts associated with the trappings and accompaniments of culture and language. If, as LaCocque suggests, life is relationship and dialogue, then humans can lose this and can twist and invert it through language in a way that animals do not. If humans alone can consciously pray then in this way too there is a spiritual death accessible only to the human. If humans alone can repent, then the lack of repentance may lead to a spiritual death in which animals do not participate. Nevertheless, when chimpanzees seek brutal conflict with other chimps, and perhaps even in predation, there is a yawning gap between the way things are and the vision of the peaceable kingdom, and in this sense all animals must participate in the spiritual death that humans experience so acutely. A part of the meaning of death in Genesis (and Rom 5) must be this spiritual dimension, which is certainly the meaning of death in John's gospel that says, "Very truly, I tell you, whoever keeps my word will never see death" (John 8:51) and can be understood as the meaning Paul was giving death in Rom 5:12. I discuss this further in Chapter 10.

And it is possible, as I discuss in that chapter, that humans have eaten symbolically from the tree of the knowledge of good and evil in a way that is distinctive and inclines us toward spiritual death and toward a fear of physical death which we may not once have had. The images in the book of Lamentations and in other prophetic works of whole societies or peoples caught up in tragic bondage and sin does reveal how humans can be enmeshed in death in ways that are unprecedented in the animal world. This is something we have experienced in recent history and in the Sudan and Afghanistan and Syria today. Nevertheless, primates too have been seen to be corporately affected by the presence or absence of social cultural constraints. A well-documented troop of baboons was affected by an outbreak of tuberculosis picked up from meat. Only the more aggressive males died and the population became female dominated. The culture of the troop then changed, with far less conspicuous aggression and lower markers for stress in both genders and more affiliative behaviors. Interestingly, the gender ratio and the culture of less aggression have persisted through several generations of adolescent males joining the troop.³⁴

How life exists at the boundary of order and chaos, and how it is that in the end an organism gives up life and gives in to death is a part of the mystery of death, but indeed we can see in the process of life two contradictory impulses at work. Thus even with death, there is the sense that what humans experience so acutely and so deliberately and so profoundly is not entirely absent from the world of animals. The same ambiguity remains; the desperate clinging onto life—the inevitable giving up of life in death. "Within the garden," says LaCocque, "there is no evil and there is no death. Without the Garden is the realm where good and evil, life and death mix. It is a vile, impure, 'impossible' mixture—and yet that is exactly what our existence in this world is made of."³⁵

Genesis, Death, and the Tragic

The close connections here described between animal and human may undermine cherished theological frameworks derived from Genesis, but this does not thereby mean that Genesis should be disregarded or rejected. There is much wisdom to be gleaned there, and much depth that ironically can aid our journey with animals if we read from a different perspective. The animal record rules out any literal interpretation of Genesis and the tight theodicy associated with that story, but not necessarily the symbolic meaning that impinges on history and science, nor the importance of history emerging out of the teleological emphasis of Genesis that LaCocque provides. Genesis, of course, is such richly deep material that it knows almost no bounds in its interpretation. Are we wrong to think that it has something to do with origins as well as the existential state of each human being? I would side partly with those who would argue that Genesis does perhaps have an historical edge, in the sense that it is concerned with origins and etiology and the meaning of evil, and it does convey an historical/tragic element, one that is then continued through the

Old Testament without ever mentioning the Adam and Eve story again. This tragedy is repeated in story after story of Israel's history and stands by the very repetition of the original tragedy to convey some meaning to us about spiritual death and about fallenness.

Genesis is not asking the questions modern people ask. It is symbolic and oblique in many ways to our contemporary consciousness, but it can be brought, nevertheless, into transversal dialogue with science. LaCocque makes most sense of this historical edge when he points to the teleological aspect of Genesis[36] and to its promise of redemption and hope, which affirms the necessity of the historical:

> Biblically speaking, evil is through and through historical. History—that alone among the nations Israel acknowledged as a dialogical discourse—is an ongoing battle with evil, that is, with moral and cosmic evils. Every partial victory over evil contributes to the liberation of the universe.[37]

But LaCocque also acknowledges that evil comes from the outside as well as from inside humanity: he speaks of the origin of evil as emanating both from outside and inside and reveals that it is always already there from either direction. The "serpent" represents the "outside" source, the human heart the "inside" source, and a twisted language is able to bridge the two poles:

> Before the human's fall, the rest of creation has already fallen. It remains true that the primal couple will eventually pull down with itself all creation to a universal trial, but the creation's fall is both *cause and consequence* of human transgression.[38]

When the burden of Genesis is not upon the redistribution of evil from God to humanity and the cognitive separation of humans made in the image of God from animals who are not, the close links between humanity and the animals become more evident. So LaCocque points to the conversation between the woman and the talking snake—an inversion of roles, an acknowledgment of their equality, a suggestion that the evil is resident first in the animal? He points to the man being the first of the animals in the J (Yahwist source) story, and the woman the last. He says this makes humans the "kings of the universe" but it also makes them animals first.[39]

The end result is not only an affirmation of history and the promise of salvation, but also a stark statement of human solidarity with all other life:

> The universality of God's love and care in J's narrative is counterbalanced by the universality of human guilt and of the "groans in travail of the creation" (Rom 8:22). Such is the lesson of J's. Implied is the conviction that the healing or redemption will also be universal or not at all.⁴⁰

If we combine these approaches it makes sense to speak of Gen 3 as referring in part to the repeated fallenness that seems to accompany each upward progression in sentient capacity; there is a suggestion of the cost or dark side that accompanies the goodness of sentience. In Genesis we see for the first time perhaps fully conscious deliberation linked closely with the goodness of full sentience, and the tragic that is to accompany all human life hereafter. It was Michael Polanyi who spoke of our eating again from the tree of the knowledge of good and evil in our technological progression.⁴¹ And just as the existential interpretation has the moment of fall repeated with each human history so it might be possible to see successive falls—successive stages in idolatry and self-will. Certainly the move to agriculture would have been such a movement, requiring as it did first collective and then individual "ownership" of land and the beginnings of a much tighter control of land than was true in hunter gathering communities.⁴² Such successive falls would certainly signify successive loss of innocence and increasing bondage, bondage to the voice of the serpent that so quixotically appears in this perfectly manifest world. Thus Genesis gives us the picture of a world gone awry, a world of perfection and corruption in which there is temptation already present. It is this aspect of temptation and of evil presence that I will pursue in Chapter 9.

> Hence retrospectively, the divine warning, "the day you eat from it you are doomed to die," (2:17) means something like: "That day, death will be everywhere. Death will be a seal on everything you do and experience. Even your science will reveal to you that all things die, it will be the knowledge of 'dead' things, as transient as you are yourselves. You will have lost all sense of immortality and eternity, trading transcendence for contingency."⁴³

All of this makes evident that while human uniqueness remains, humans share with animals in an overall history that includes the tragic. The higher animals have the capacity for suffering that is significant. Animals living in stable communities nevertheless eat their young and are aggressive and violent to others; they set out to annihilate their neighbors. They are often vicious and deliberate predators. Yet they also exhibit the love, companionship, fellow feeling, empathy, and compassion that are the precursors of all we value in our common life together.

We now strongly suspect that humans did not find themselves entirely innocent at the boundary between species. Wherever the species line is drawn inevitably our forefathers and foremothers came to being human with propensities for both love and hatred. There was no paradise, no line of demarcation drawn in the sand. This dismantles the kernel of any literalist approach to theology, but also any more ambiguous use of Adam or fallenness or any of the categories that dig deep into human consciousness and the logic of Christian systematic theology and Christian liturgy. The type of fallenness of which I speak was not a fall that invented death or suffering or even wilfulness. At the edge of becoming human our newly powerful minds were capable of new heights of both goodness and evil; the threshold of humanity brought with it also new depths of moral accountability and remorse. But this was not a fallenness that made the difference between good and evil, death and paradise.

In this context the Amazonian Piraha tribe is of enormous interest. They do not share many of the cultural traits we take for granted. They are without material culture, without numbers, without a sense of history or of the distant future; they have no God, no leader, no creation myth, no violence amongst themselves, no mimetic desire, no material culture—and no felt need for anything outside themselves. They are content, with a unique language almost impossible to learn. This people make us wonder about the nature of the knowledge of good and evil, and the extent to which the seeds of fallenness—mimetic desire and a tendency to aggression—might be relatively absent in a human population while seemingly present to some degree in nonhuman ones.[44] The existence of hunter-gatherer groups like the Piraha should make us less certain of exactly what it is that defines the essence of humanity's undoubted uniqueness.

Because we humans tend to see ourselves as so different from the animals, we do not own the "inheritance from animals" thesis. Moreover, humans are afraid that owning or acknowledging this history would detract from all notions of responsibility and moral categories. I would argue, though it is not the main import of this monograph, that indeed human/animal kinship does shake up but does not annihilate our categories of what it means to be responsible and free. Inevitably it means that animals to some extent share in the moral life and humans are more bound by inheritance than we once supposed.

This boundedness also means that our sinfulness and our moral natures do not necessarily depend upon acts being done in states of radical freedom. This has long been acknowledged in reformed theology from Augustine to Luther to Calvin to Jonathan Edwards and to Schleiermacher. Freedom is not the freedom to will as we will, only the freedom to act as we wish. This wishing and willing has a long history, a history we know goes deep into our animal past. That we feel responsible even when in states of relative innocence is evidenced by survivor guilt and by guilt that is felt when a person accidentally kills or when people kill in a state of psychosis, or as part of an authorized group such as an army. Recognition of this shared propensity to sinfulness and the deep roots into our animal past could make us more compassionate of others and more focused on redemption and renewal rather than punishment. After all, only humans have access to the story that makes redemption possible, which might lift us out of this quagmire of inevitability and into the possibility of forgiveness. If there is human uniqueness it is in the sharing of the possibility of forgiveness and not in violence, sin, or death.[45]

Conclusion

Thus it is no accident that the particular anxiety that overtakes students when evolution and our kinship with animals is raised has to do with death. Death is central to the logic of Christian theology, and it holds a particular existential force for us as Western, technological socially isolated people, even for people who do not believe Christianity is associated with the "solution" to the problem of death. Perhaps for this reason evolutionary theory with its obvious denial that death is uniquely linked to humanity is seen as the alternative to belief. Perhaps also this is why problems of theodicy are particularly noted by the opponents of theology.

Death and fall are the kernel of Christian theology and the foil of salvation and grace and have traditionally been seen as the justification of the separation between humanity and animals. As evolutionary theory exposes this false interpretation we also see for the first time how close we are to animals and how perhaps God has always intended we should see things this way. Even the sting of death is something that animals can be said to share to some degree. With the loss of the grammar of a single and dramatic fall, however, we also lose a powerful explanation for evil. Where do we look when this is gone? How do we understand evil in dialogue with science? We already find in Genesis powerful rhetoric that underlines solidarity between human and animal and indeed all life. There are also hints that for humans and animals together sin, death, and temptation come as intrusions and disturbances. In the next chapter I examine some of the ways theologians and philosophers have dealt with the problem of evil for humanity, animals, and for matters of evolutionary suffering.

3
Animal Suffering—Philosophical Responses

> *I own that I cannot see as plainly as others do, and as I should wish to do, evidence of design and beneficence on all sides of us. There seems to me too much misery in the world. I cannot persuade myself that a beneficent and omnipotent God would have designedly created the Ichneumonidae with the express intention of their feeding within the living bodies of Caterpillars, or that a cat should play with mice.*
>
> CHARLES DARWIN

I ARGUED IN the first chapter that our knowledge of animals exacerbates the traditional theodicy problem because the strict threshold between animals and humanity is broken down; animals are capable of acts of proto-moral aggression while some human communities live in relative peace. Humans undoubtedly came to the species boundary with a propensity to sin, as we might imagine chimps or bonobos would be similarly afflicted were they to achieve full sentience and sapience. There never was a boundary between paradise and the fallen state, thus the old story inspired by Gen 1–3 and the old logic that helped explain evil no longer works. In biology, a hyper-Darwinian evolutionary paradigm that assumes that the *only major* driver of evolution is gene mutation and natural selection at an individual level intensifies the predicament even further because evolution appears to be both random and uncaring while being the responsibility of the Creator alone, if God exists at all. Or so it appears; the evolutionary process as it has often been described is fueled by a sometimes cruel competition for scarce resources that can only be seen as counter to Jesus' ethic of love. Thus theology is left without the traditional defense of God and with a God who perhaps created by means that were inconsistent even with the human requirement for ethical behavior.

Evil still cries out for some explanation. One consequence of the rhetoric of the hyper-Darwinists has been to undermine deep human religious impulses. These impulses have always been generated by an intuitive grasp of the beauty and complexity of the natural world. They have led to belief in and worship of God long before Scriptures existed. In a post-Darwinian world piety is undermined because beauty has been explained away as inherently subjective and accidental, as spandrels (order that is achieved as a side-effect of other processes), or as self-realization of a remarkable but impersonal process.[1] We need some way of explaining and understanding why beauty and order are mixed with tragedy and disease and suffering. A literal approach to Genesis, or at least a literal space-time fall, fulfilled that function, and theologians and the faithful in the past were confident of drawing links between natural beauty and divine love; the beauty was obviously God's influence, and the evil was all the work of the Evil One or the consequence of human sin or punishment for sin. Before the nineteenth century humans focused on the human time span and not on the long prehuman history of evolutionary becoming. This chapter examines the possible philosophical explanations for evil in light of current awareness of animals and their long evolutionary history.

Thus I will look at the state of play—insofar as that can be done concisely—in philosophical defenses of God and arguments against God's existence in light of what we know of animals and of evolutionary history. Those seeking deeper interactions at this stage should consult Michael Murray's *Nature Red in Tooth and Claw*. In this chapter I assume that there is evil that is horrendous enough that it cannot be dismissed as merely pain that is a normal part of an animal's growing and defense against predators and sickness, even if we might all place the locus of such suffering in slightly different places.[2] I assume also that the neo-Cartesians are in error, as I have argued in Chapter 1, and that animals do truly experience morally significant suffering, even if that suffering never reaches the heights of self-conscious human misery. I assume, moreover, that there is pain and suffering that goes beyond what could reasonably be considered to be necessary for character development—or even soul making—in either animal or human.[3] And I assume that a good God cannot be defended with reasons that seem to undermine God's very goodness. Thus arguments that suggest that evil at the animal level can be excused by the great value of human nature it produces are not satisfactory defenses of the goodness of God.

Natural and Moral Evil Are Not Distinct

Philosophers have defined natural evil as that suffering that is caused by so-called natural calamities such as earthquakes and tsunamis, drownings, and fire. Moral evil is evil where responsibility can easily be placed at the hands of moral agents or human beings. Thus the Holocaust and murder and rape are all examples of moral evil. But the lines become blurred, especially when we consider evil on an evolutionary scale. By everyday standards it is natural evil that is hardest to understand.[4] When considered within the vast range of evolutionary history, natural evil is enormous compared to anything humans might do. The more we know about animals the more it seems that they suffer, though they do so in states of relative innocence. The more we know about higher animals, the more it appears that they deliberately harm at a proto-moral level, with intentionality and limited planning though without the self-reflection and ideological and social embeddedness that humans express. Thus suffering has abounded over at least hundreds of thousands of years, as has the harm done by these higher creatures. Behind all of these instances of suffering God is the one who stands in the dock, accused either of nonexistence, cruelty, or indifference.[5]

The turn of attention to evolutionary history makes evident to our consciousness the long line of creatures existing in the shadowlands between complete innocence and human culpability. Therefore, humans reached the species boundary already with proclivities to sin. Although humans might be responsible at one level for what we do, at another, God lies behind our very becoming in the form in which we are. Thus moral and natural evil blur at points, all being seen in some sense as varieties of natural evil of which God might in some sense stand accused. While in the past many evils were thought to be the result of the fall of humanity and humans were ultimately responsible, now the dice have been cast in the opposite direction, and all evil can in some sense be placed at the feet of God.[6]

Predation

Apart from the fuzziness surrounding the prehuman/human boundary there is also the problem of predation. Looking back at the evolutionary process we see hundreds of millions of years of predation, competition, and survival of the fittest against all odds. And although I will argue that other mechanisms and proclivities are also involved, predation has characterized the evolutionary process from the beginning. Why has God let all that happen only to turn around and tell us that the mystery of love is at the heart of reality? Why is God working, or appearing to work, in essentially a Darwinian fashion? Do the ends justify the means for us, as perhaps they do for God?[7]

What we have in the blurred boundary between ourselves and other animals, and in the long history of ingrained killing, is a more complex, more brittle problem of evil. Evolutionary history negates the possibility of an historic fall as we once knew it. There never was a time of paradise from which human and animal have fallen. It is sufficiently brittle that Richard Dawkins's and Sam Harris's cynical rejection of faith can appear quite plausible. How can we begin to resolve this issue?

Christians believe that the creation is essentially good and is the creation of a good God—though we will discuss this belief further in this chapter. At least we know that creation is not a distraction from the good or a plaything for the gods, as it is in some ancient or Eastern religions. For this reason natural disasters offend us. And for Christians the point of life cannot be evaluated merely in terms of an afterlife. This earth, this world, this cosmos are all good. The creation has always been understood as "God breathed," full of Spirit and *logos*. God's Spirit has been understood as indwelling the world continuously. God's *logos* is the wisdom, the principle, the inmost being of the cosmos. We feel betrayed when nature, which has nurtured life, becomes life-defeating for us, even if the shearing of tectonic plates in an earthquake will ensure the continuity of life in the distant future. After all, Jesus has assured us that God knows the fall of the sparrow and has provided for the lilies of the field.[8] The indwelling presence of God in creation for the Christian theist makes even more acute the problem of evil than it is for the philosophical theist.

The Logical Argument from Evil

Who is God and how can God be defined? This has always been a difficult question. The omni-God of the philosophers is an omnipotent, omniscient, and perfectly good God. Such a God should be incompatible with evil, especially the extent of the evil we know exists and has existed. In the introduction I argued that Old Testament scholars will claim that the

God of Scriptures is more than the philosophical stereotype, or that there are significantly subtle demarcations in the way God exercises omnipotence and omniscience and that we must take care in judging what is good and what is evil. I have also hinted that theologians want to differentiate God from a mere Producer or Cause of all that is. Nevertheless, the grammar of prayer and the experience of many faithful do connote a God who participates to some extent in the practice of powerful agency. Thus the omni-God, although possibly "wrong" at the core, is not a bad bare-bones beginning for ponderings about evil. Any God who has created the heavens and the earth must approach omnipotence and omniscience. Thus it is worth considering the arguments against this omni-God. I would contend that nuanced though the God of Christian orthodoxy might be, the attributes of knowledge, power, and goodness cannot easily be denied in the God revealed in the Scriptures, even if this God's exercise of power is ultimately a subversive power in weakness or a dynamic interaction within the Trinity and with created beings.

The logical argument from evil purports to show that it is logically inconsistent to posit an omni-God who can remove evil but doesn't. Nevertheless, there are what is known as defenses, or speculative theodicies or explanations for God's allowing evil. These are philosophical or theological arguments that give some sort of explanation of how it is that evil might coexist with an omni-God. These arguments are not necessarily compelling, but if they are successful they may be the means of acquitting God of nonexistence or malice.

The most common defense of the omni-God in the face of evil is some form of the free-will argument. This defense argues that the freedom of humans demands that God not be always intervening. Parents understand this. We cannot rule and manage our children's lives beyond an age where that might be appropriate. Even young children require some measure of freedom. Moreover, many have been prepared to extend this argument to other creatures, to make a "free creatures" defense, fulfilling the requirement that all of this makes sense at the level of animals as well. At all levels of sentience some measure of freedom—and danger—is required.

Not surprising, this defense, although it has some truth in it and is most often referred to in popular defenses of God, is insufficient in itself. In a recent article John Bishop and Ken Perszyk argue that if God is perfect, God could have "actualized a world containing free creatures but no moral evil just by selecting those initial conditions and strict deterministic laws of nature which together would entail that every free creature always, but freely, chose the good."⁹ After all, is this not what heaven will be like? There is a counter to the argument of Bishop and Perszyk in the famous Plantinga defense. In Plantinga's argument God has so-called "middle knowledge"; God knows all the possibilities that ensue from free creatures, but not what each creature will actually do. In fact, Plantinga explains, it may be the case that in every possible world some agent will commit sin. Thus he claims to have invented a scenario by which in a world of free creatures evil may get out of even an omni-God's control.¹⁰

Bishop and Perszyk doubt, however, whether a God who knew that freedom would bring about horrendous evil *should* have created free creatures. One might ask whether anyone has the right to make such an assumption of course. At times like this philosophers and theologians differ. But they also argue that even if God were thinking ahead to the great value of free creatures, is this not utilitarian in a way that God perhaps should not be, especially for the animals—and many of the humans—who were not to benefit? Moreover, with speculative theodicies such as these, Bishop and Perszyk caution that in all cases God upholds and ensures the conditions under which evil is able to flourish, only then to turn around and redeem it.¹¹ This apparent conflict casts doubt on the omni-God's existence or goodness. Nevertheless, humans gain some measure of reassurance by the free will and other rationalizations of evil. Animals have no access to such conceptions. Of course animals have no need of reassurance as far as we know. As the moral observers of the universe, however, humans need some way of understanding animal suffering as more than just a means to a higher order of different creatures. That state of affairs would be so inconsistent with all else we know of the God of Scripture.

Moreover, there are many human situations that seem to be puzzling, even within the bounds of the free-will defense, because free will, if it is really so valuable, should be able to be freely relinquished at times. In human relationships we give ourselves into the care of others at times or participate in institutions that require some curtailing of our freedom. Isn't prayer such a voluntary relinquishing of freedom relative to God? Recently in New Zealand, for instance, seven young teenagers from a Christian school were drowned while canyoning on a school trip. There was a great deal of prayer going on before the students left. They went into the canyon on their last segment with special prayer. Escape routes and solutions existed but were not seen or taken. If prayer is anything it is surely giving up rights to autonomy on this particular stretch. No freedom would be impinged if a loving God had put hurdles in the way of this particular enterprise, or had given inspiration to someone to see a way out; and there are numerous stories of such miraculous escape through a voice or inner thought or dream in the past. Some people do report sudden and supernatural responses to prayer; they are saved miraculously. An angel whispered in Joseph's ear more than once, and this biblical pattern has been reported constantly by believing Christians throughout the history of Christianity.[12] The freedom of these people was not compromised, as it would not be if a loving parent intervenes to save a child. For animals also, the experience of suffering and affliction is uneven and random.

Nevertheless, others argue that if God were to interfere even once that would compel God to intervene at all times. God must not play favorites, nor can prayer be used as a transaction with God. Yet full consistency does not work well in theology, for Jesus healed and did miracles when asked, and yet this was plainly favoring only a very few suffering souls in a great ocean of early first-century affliction. The language of "intervention" or even of agency does not, of course, really work for God; nevertheless there are problems at the level of justice, personal piety, prayer, and logical coherence surrounding some of these events.

The second major category of defense is the "best of all possible worlds" argument. There are many variations of this argument but all assume that somehow in ways we do not understand, the world could not be anything but this way—especially if free creatures like ourselves were to evolve. The "free creatures response" thus assumes a "best of all possible worlds" argument. This argument does depend upon the very uncertain assumption that there is no other way, or that other ways have worse outcomes. There is probably truth to this defense also, especially when we get to the level of asking whether human life in the way we know it doesn't require this level of danger for knowing of good and evil. But again the distribution of pain is very uneven among humans and among animals. If evolution can produce peaceful colonies like those of bonobos and cooperative (if sometimes brutal) ones like those of meerkats, as well as good parental models like those of elephants, why are animals and humans not all like that? Animals differ greatly in terms of their level of predation and aggression. There do seem to be more cooperative, less aggressive models possible. A similar situation exists with humans. Moreover, if anything is responsible for the high level of aggression in all species it might be the ruthlessness of natural selection, always producing animals that are capable of taking the upper hand, and here God must be responsible. Is there no other way that evolution can happen? There is certainly enough uncertainty about the truth of this being the best of all possible worlds to motivate more efforts to understand.

As theologians we often rush into our well-developed theological defenses, sometimes with echoes of the Adamic theodicy lurking in the background, without taking full cognizance of the severity of the philosophical arguments against the omni-God's existence. This situation is all the more serious for theology because, although theology has greater resources than philosophy, people of faith also have greater need of reassurance of God's presence. Plantinga's defense, for instance, although it might defeat the logical argument in some way if Bishop and Perszyk turn out to be wrong, is nevertheless not of great use in Christian theology. How can we argue that a God who has long been thought to be revealed in Word and in Christ and through the Spirit in experience and nature can only, in fact, be defended by an obscure argument that very few of the world's human inhabitants could ever understand? Michael Murray admits as much when he says: "citing certain possible, for-all-we-know reasons for the reality of seemingly gratuitous evil does not remove the persuasive force those evils have; and this leads us to doubt the existence of God."[13] How can we argue that although the omni-God probably does not exist, some very slightly different God—the God of Scripture—does exist? The difference is a fine line indeed. Moreover, the logical arguments apart, there is also the evidential argument against God, which claims that given the extent of evil, God probably does not exist. Yet faith itself requires evidence, strong evidence really, that God *does* exist. We cannot with integrity always be saying that yes the odds do seem to be packed against God and the evidence does not look good, and yes there is the possibility of a clever trick of reason that might save God, but we will all believe in spite of the evidence. In the end that would be foolish, and it would certainly not be respectful of truth. At the very least only the very few virtuosos of faith would accomplish this—those whose certainty of personal experience outweighs all intellectual scruples.

The Evidential Argument from Evil

I present here only a very cursory summary of Michael Murray's very detailed book, *Nature Red in Tooth and Claw*, which is an attempt to provide a variety of CDs (*causa Dei*) or explanations for evil in spite of the problem of animal suffering. Murray surveys a number of CDs, any one or combination of which might be used to defend God against the charge of nonexistence or imperfection given animal suffering. One CD will always be that we have not the cognitive capacities to understand these mysteries. But if we are arguing for God against substantial evidence we would do well to consider the perils of self-deception and look further. Murray begins with the Neo-Cartesian argument that animals don't really suffer. He judges this to be valid, our never being fully able to plumb the depths of animal consciousness, but he concedes that few would now be convinced by it.[14] Murray next considers the traditional fall and Satan as explanations for evil and grants the consistency of this approach, but he is cognizant, again, that few in the West would any longer judge a Satanic explanation as likely.[15] I would argue here that Murray builds up a case of a straw man Satan and easily pushes evil aside. Evil, however it is described or personified, cannot be a simple explanation for sore backs or avalanches; rather it must be a subtle influence—hidden and elusive in its ontology and agency.

The demands of a greater good like freedom are considered as a possible CD, though in this case animals are prone to be seen as suffering for the cause of humanity. Murray considers also the theological responses normally attributed to humans—resurrection and soul making—and admits that the first of these is certainly one possible CD. The most interesting CDs in a sense are those that look at nomic regularity. This is the idea of the lawful regularity of the world that is required for anything to be the way it is now. There is something about the values of freedom and choice and moral goodness, argues Murray, that might require that the whole cosmos be ordered in the regular and predictable way it is now and to have the history it has had. He extends this to the idea of a "chaos to order" universe, that only out of certain levels of chaos and discord can any value emerge, and he claims that this finally is the most persuasive of CDs.[16] Evolution must go through the stages it has done, and in order to have human beings with sentience, sapience, sensitivity, and ability to suffer, animals too must show the beginnings of these signs:

> ...we would expect to find, in our ancestral history and indeed among our contemporaries, organisms with many but not all of the mental capabilities possessed by humans. Forms of mentality, including the ability to experience pain and suffering, would thus likely be found among these precursors.[17]

In the end Murray argues most strongly for this "chaos to order" universe, which is a modified argument from nomic regularity. There is something about the universe, he says, which requires at least pockets of order which have progressed from chaos. Such nomic regularity also requires an animal prehistory with attendant suffering. Although Murray is defending God, his book makes plain that if we accept the suffering of animals, this defense is not easy. He draws out the ways in which theology has previously leaned heavily on the Adamic Fall and notions of human free will. In the end, Murray's "chaos to order" universe, although convincing at one level, is less so at another because if God is to be the object of worship then God must be evident in the here and now, in the suffering and the pain and the numinous, and not just in the overall picture, useful though such a picture is in the life of faith.

At this point it is worth acknowledging the ways in which evolutionary theory and the carnage and predation involved through millions of years predating humans or even hominids have enormously complicated and deepened both the evidential argument against the omni-God and in fact the logical argument. The evidential argument is augmented by the relentless carnage and animal suffering that has gone on in times when humans could not be even slightly to blame—unless one is accepting backward causation. God is at least implicated in all that comes under the heading of the evolution before the rise of hominids. If we consider only human misfortune such as sickness or natural evils in the human era, there is always a distancing from God that is possible. Moreover, as I have indicated before, the evolutionary timescale makes impossible the most significant speculative theodicy, that in Adam the whole earth has been cursed and broken.

Contemporary Speculative Theodicies

In the next chapter I look at contemporary speculative theodicies. How has theology faced the challenge of the argument against God, bearing in mind the point of Bishop and Perszyk that it is not good enough to postulate that God redeems what God has created if the values God reveals in creation are completely opposed to those shown in redemption and at odds with the revealed God's goodness. If it were not for the hardness and callousness of natural selection we might easily be satisfied with redemption as an answer if it includes the whole ecosphere; though even there, faith has always insisted on there being intimations of redemption—resonances of the Spirit in this world as well as the next. For most of us that probably seems like a plausible enough deal, and we might easily assume by faith that it is in fact a good deal for those who do not see it and for animals that can make no judgment. But there remains the argument: why create in this cruel way to start with? Why not make a world of lambs and kindly straw-eating lions, of sleek but agreeable cheetahs? Does it have to be a world of brutal predation and holocaust? Is it reasonable to suggest that love is the highest value in the universe but God could not make this a reality at lower levels of sentience, even when the demands of being free were relatively less taxing, as they must be for nonhuman animals?

Conclusion

The philosophical arguments I have been exploring for the existence of God given the presence of evil can in some sense be largely included under one heading: in spite of evil, a greater good will be the result or is the case. Thus some argue that the high freedom of humans requires the presence of evil; others argue that a greater good as yet unknown to us will come of all the evil. Together with this argument is the argument that this world is the best of all possible worlds. Thus philosophers, if they are so inclined, argue that a case can be made for the existence of an omni-God—this *might* be the best possible world, there *might* be values we would all agree are worthy in spite of the evil, it *might* be the case that freedom demands all this and thus it *might* be the case that there is no gratuitous evil. It *might* be the case that animals had to suffer for hundreds of millions of years so that humans could understand these things. It *might* be reasonable to attribute evil to kenosis, or the absence of God, and then, when we see beauty and love, to think "this is the providence of God." Thus belief in God in spite of evil is rational, though some would still say it was unlikely. It can appear unlikely because the evil is so unevenly distributed and because, on the surface, there appear to be ways that God might have done things differently even though this assumption can also be regarded as arrogant. I would argue that when we look at the extent of evolutionary suffering and the magnitude of animal suffering, this unlikeliness of an omni-God's existence is only increased.

Is there then a way forward? Karen Kilby has argued that Christian theology has two options. One is to be philosophically consistent and to end up with a God who is either not all-powerful or nonexistent—and this has for the large part been the path traveled in the twentieth century. The other option is to admit the biblical revelation of God and to live with the seeming inconsistency or paradox surrounding evil, finding in the mystery of the path of salvation a consolation that is beyond our understanding. This way we affirm that we cannot find an answer even though the questions themselves are valid.[18] This living with paradox does not come easily to us as children of the Enlightenment, but it does square most easily with the revealed God of Scripture. We might ask, however, is the puzzle over evil a paradox or an error—a contradiction? And living with paradox is done rationally only if the goodness of God is overwhelmingly evident in spite of the evil. Moreover, why accept paradox and mystery when the atheists claim to have a more consistent and more plausible explanation for all that is? I will argue in Chapter 6 that affirming the biblical parable of the wheat and the tares is one step toward trying to show how we can make coherent sense of the mystery, even if we do not understand it. In the next two chapters I examine some of theology's responses to intractable evil and the question marks it places over our understandings of God.

4
Animal Suffering—Theological Responses

No, it won't do, my sweet theologians.
Desire will not save the morality of God.
If he created beings able to choose between good and evil,
And they chose, and the world lies in iniquity,
Nevertheless, there is pain, and the underserved torture of
 creatures
Which would find its explanation only by assuming
The existence of an archetypal Paradise
And a pre-human downfall so grave
That the world of matter received its shape from diabolic
 power

C. MILOSZ

Denying the Omni-God

At this stage it is proper to ask what kind of God stands accused. The omni-God of the philosophers is the God who is powerful enough to stop evil, all-knowing enough to anticipate evil, and perfectly good and loving in character and intentions. The problem of evil looks at the ways in which this good all-powerful, all-knowing, and loving God might be compatible with the evil we see around us. If God is good whence comes evil? If God is all powerful cannot God remove evil where it appears? If God is omniscient there is no excuse for God's not knowing that the evil is or is about to happen. In the face of this argument and the manifest degree of suffering around us, many give up on the idea of the omni-God as too inconsistent with the facts.

Deism

Thus we have both process theology and deism. Deism is more often assumed than argued. For many scientists this is the default position, blending as they sometimes do their methodological naturalism with a more metaphysical kind. Deism has conceded too much to science—and the authority of the Central Dogma—and has pushed God's influence out of our sights. In deism, though, God does not intervene because God cannot. This is not the type of God God is. God is all powerful in terms of beginning the whole thing, but is inaccessible and nonintervening. Because God is so inaccessible, belief in this kind of God and an atheist's belief in the universe itself become inextricably intertwined. The deistic solution does a great deal for consistency but nothing at all to satisfy the needs of piety and the ubiquitous experience of the Spirit of God (especially in the twentieth century church) or the narrative of Scripture, which refers constantly to an active and immanent and powerful God.

Process Theology

Process theology, however, has made God ubiquitous as lure or persuasion but only the observable equivalent of a weak force that might not in the end persuade.[1] Process has been widely appropriated in the twentieth century, solving as it does both the problem of evil and the problem of a patriarchal God, as well as the need for God to be hidden. Even some conservatives have come very close to process in the debate over the openness of God.[2] In process, all units of being, from the atom or cell up to self-conscious human beings, have a measure of awareness or "prehension" of all impinging influences. This feeling is generated not just by means of external efficient causality but, what is more important, also at a subjective inner level. Prehension increases from being almost nonexistent in the atom to the unique subjectivity of the human consciousness.

The other important characteristic of process is that identity is maintained over time, not by some dualistic essentialist properties of soul or spirit, but in a series of momentary "actual occasions" or events—entities are in a continuous process of creation and perishing. Future possibilities are made known to the occasion by God who "lures" but does not force each occasion toward greater complexity and harmony. God is the source of all values, understanding the most advantageous future for all entities.

The energy released as the concrescence is achieved propels the occasion into the future, where it in turn becomes an object for further occasions. Thus identity is communal and corporately derived, reflecting the participation of the totality of people and other units in an entity's past and possible future. Subjective immortality is achieved by each actual occasion as it is prehended by and incorporated into God.

The past and the future are the domain of two different poles; all reality from the most basic unit to the highest, God, is dipolar, constituting a physical pole and a mental pole. The physical pole is not, strictly speaking, something material, but it is the aspect of the unit that prehends or feels the past as object or given. The mental pole harmonizes the past in conjunction with an understanding of possibility in the future. God also has two poles, the primordial and the consequential. God perfectly prehends the whole world and moves to participate in the created world—or divine material pole—via the initial aims or divine lures in each actual occasion. Although the process God is eternal, transcendent, and necessary in the primordial pole, this God is also limited and contingent in the consequential pole. The future is thus open to God as it is to us.

Process does affirm, with traditional theology, a purpose hidden in the basic structure of the universe. It affirms the interconnection and interdependence of all living and nonliving things and has therefore also been popular in ecological theology. The process God, however, is a departure from the omni-God because in process theology, as in Deism, God cannot intervene because God is not that kind of God. Moreover, even omniscience is given up. God does not know the future until it has happened—as is also the case for "openness of God" theology.

Both process and deism are entirely consistent and solve the problem of evil. Process has become the all but inevitable final position of those who delve deeply into the science/theology dialogue. I would argue that both deism and process may be seen as attempts to foreclose on integration of the scientific story and the theological story too soon. The process and deistic solutions attempt to understand God, science, God's interaction with the world, and the presence of evil in a coherent manner. They both allow some settling of the problems that arise in dialogue between the concerns of science and faith, but only if other issues are overlooked. What have been overlooked very often are the biblical text and the experience of believers. Faithful people all over the world interact with God and report miracles, healing, and answers to prayer—intimations of a more dynamic relationship between God and the world. These cannot be easily dismissed or reinterpreted as lure. The power of God is discernible, as it was for Job, though it is not distributed in a way we would like or would even deem fair—it is experienced more as gift and hiddenness than as law.

Orthodox Solutions: Kenosis, Imperfection, and Fully Gifted Universe, and NIODA

Kenosis used to be a theological insight which tried to solve the problem of the two natures of Christ and of Chalcedonian unity. God divested Godself of divine attributes—the relatively unnecessary ones—to take on human form. The idea of kenosis, however, has been revised with profligate abandon to refer to the giving up of Godself that the very act of creation entails. God must make room for creation and in order so to do God removes Godself from the space of the universe, voluntarily giving up freedom and control.[3] Surely, it is argued, the creation of something outside of God or other than God is an act of extended kenosis, a limiting of God's powers in generous love. In this way there have been attempts to reconcile evolutionary history with a more orthodox Christian faith—although sometimes in combination with a process perspective.[4]

Moltmann makes links to the Jewish concept of God's *shekinah* (the visible glory of God's presence), which dwelt with the Hebrews in their humiliation.[5] He takes up Isaac Luria's evocation of the infinite Holy One, "whose light primordially filled the whole universe, who withdrew his light and concentrated it wholly on his own substance, thereby creating empty space."[6] God suffers alongside God's people:

> How can a finite world co-exist with the infinite God? Does it set a limit to the limitless God, or does God limit Godself? If this limit or frontier between infinity and finitude is already "fore-given" to God, then God is not infinite. If God is in his very essence infinite, then any such limit or frontier exists only through his *self-limitation*. That makes it possible for a finite world to co-exist with God... Only God can limit God.[7]

Moltmann argues that "'The Lamb slain from the foundation of the world' (Rev 13:8) is a symbol showing that there was already a cross in the heart of God before the world was created and before Christ was crucified on Golgotha."[8] He goes on to say that if kenosis applies to creation, it applies even more to preservation and consummation.[9]

God acts in the history of nature and human beings through his *patient and silent presence*, through which he gives those he has created space to unfold, time to develop, and power for their own movement. We look in vain for God in the history of nature or in human history if what we are looking for are *merely special divine interventions*. Is it not much more that God waits and awaits, that—as process theology rightly says—he "experiences" the history of the world and human beings, that he is "patient and abounding in steadfast love," as Psalm 103.8 puts it?¹⁰

Here kenosis and process come very close, as Moltmann himself says, and there is a flattening out of divine activity to be merely or only the silent presence. Associated with kenosis is the Moltmannian emphasis upon solidarity and the God who suffers with humanity; in fact for Moltmann the kenosis of the creating and preserving God is not the abandonment of the creation by God. Kenosis points to the future and to the purpose within the creation one might expect if God's spirit is immanent within it. "The goal of God's *kenosis* in the creation and preservation of the world is that *future* which we trace out with the symbols of the kingdom of God and the new creation, or 'world without end.'"¹¹ God in Christ suffers with all that suffers, human and non-human alike. The bringing to light of the incarnation and the emphasizing of this central tenet of Christian faith is a huge step theologically in the problem of evil, though we return again and again to the vision of a God who is only silent and only waits. But this is all less helpful for non-human creatures. In what way may it be said that they benefit from this solidarity with God? Humans can contemplate the mysteries of this momentous event and be comforted. The lamb or giraffe torn apart by a horde of lions has no such comfort that makes sense to us, at least. Yet God may be with the suffering creation in ways unknown to us. Yes, there is still the question, as Bishop and Perszyk argue, about why God supports the pain and suffering inherent in natural selection only to be finally in solidarity with it and redeem it too.

John Haught also points to an expanded kenosis: God both suffers with the creation and lets creation be, making room for creation. Thus evil is experienced as the creation is making its way, sometimes losing its way in this point of separation from God.¹² Evil is the experience of the withdrawal of God, which is required for life but is not ever the final Word of God into life, whether human or nonhuman. For Haught evil is also necessary because life which is other than God cannot be made perfect altogether and all at once.¹³ Thus for Haught and for Moltmann, kenosis is an explanation for the sense of the absence of God, which is, nevertheless, the *suffering presence* of God.

On one level Moltmann and Haught must be right—their arguments accord with the hiddenness of God. But at another level God's silent presence is not enough, and humans often praise God for a much more direct sense of God's presence. Moreover, there is no evidence of increasing perfection; both the wheat and the tares are growing in history. Evil cannot be understood only as the birth pangs of sentient becoming. It is true, too, that if God is praised when blessings flow and is also thought to be silently present in suffering, this can be confusing. Elizabeth Johnson says of God's actions in history that they cannot be understood to be completely hidden and obscure.

> When received in a faith context, historical research can indeed strengthen as well as challenge faith, for divine presence and action in the world are not so intangible as to leave no discernible historical traces.¹⁴

The same can be said for God's action in creation.¹⁵ If we are left with a situation where science can put up a very good case for God's not being needed as a hypothesis and genuine experience of God is thought to be illusory and death is everywhere ubiquitous, then faith can be a very grim believing in spite of the evidence. A kenotic answer to evil is also not easy to understand, not nearly as easy as the Adamic theodicy is. I would argue nevertheless that we must be prepared to identify the signs and traces of God's presence in the world as well. Indeed, Moltmann does make this argument, pointing to the accumulated wisdom inherent in the DNA and the culture of all creatures.¹⁶ Haught points to the "deeper than Darwin" aspects of the creation.¹⁷ These I would wish to highlight as much as kenosis. The experiential evidence for God is of the utmost importance in surviving the depths of evil.

Intratrinitarian Kenosis

A similar more recent theory of imperfection is to be found in Christopher Southgate's intriguing book, *The Groaning of Creation*.¹⁸ Southgate also finds need to explain the long suffering of animals and ultimately argues for their inclusion in the renewed order. He is unwilling, however, to grant the need for the giving up of *space* by God in creation, something Karen Kilby has also argued.[19] At the heart of *The Groaning* is what Southgate calls his most speculative chapter on intratrinitarian kenosis, where he makes the move that most modifies the "this is the only way" response. And indeed this chapter is a model of imaginative theologizing, based as it is on the premise that things as they are were meant to be that way. Traditional kenosis arguments, as described above, are used to explain evil as a result of God's having given up aspects of power—or withdrawn from areas of divine space—in the very act of creation, in order to let the universe be, while nevertheless coming to be with the creation in suffering solidarity.[20] The resulting reality is of course very close to process theology—as Moltmann suggests. Southgate objects to this on the basis that we have no reason to believe that the spatial metaphor applies to God. Why should God need to give up space for creatures to be?

He agrees with Moltmann, however, in his basic move of extending the divine love and sacrificial suffering of the cross into the very nature of God. "The self-abandoning love of the Father in begetting the Son establishes an otherness that enables God's creatures to be 'selves.'"[21] The mystery of imperfection and ambiguity thus lies in the unfinished nature of being a self. Each individual creature has "existence, pattern, and particularity." These are the gifts of the Logos and the Spirit. The Logos may affect the movement and dimensions of fitness landscapes; the Spirit gives particularity. This coming to be according to a pattern and growing into particularity is a part of the "selving" process, a term he borrows from Gerard Manley Hopkins. Creatures "selve" successfully when they flourish as they were meant to be and God "encourages" them in their exploration of fitness landscapes; however, in this world the flourishing of one creature is often at another's expense, and all creatures begin their lives by self-assertion.

From this self-assertion and conflicting flourishing stem all the ambiguities of the evolutionary process. The Spirit makes possible the exploration of other ways of being: of being in community and ultimately of self-transcendence, which takes an animal from self-assertion to flourishing and praise. This happens by the very power of selving that stems from the deep intratrinitarian movement of the Son from the Father and back in praise through the Spirit to the Father. For humans it is possible to go beyond this selfish level of selving to transcend the self and thereby to move to a position of praise. Animals do not often transcend themselves and many humans also fail, leaving levels of the less than perfect interaction. Southgate says:

> Theologically we may posit that the frustration of the creature, be it of the insurance pelican chick, or the sheep parasitized by the worm *Redia*...is received by the Son through the brooding immanence of the Spirit, and uttered in that Spirit as a song of lament to the Father. All that the frustrated creature suffers, and all it might have been but for frustration, is retained in the memory of the Trinity.[22]

Selving is a very attractive idea, as is the notion of solving the theodicy problem by looking to the nature of the Godhead, including the roles of Logos and Spirit. But why is selving a process that produces pain? Why do we get hints of this pain even in the Godhead? If the three persons of the trinity are in states of distinction but nevertheless of perfect love, why should selving be so problematic? True, there is such a thing as moral development in humans, who are after all finite creatures. And feminists have long spoken of the need for a level of selfhood before any notions of sacrifice might be countenanced. But why has God produced creatures who have such trouble in fully transcending themselves? Why do some creatures exhibit high levels of altruism while others have a built in mechanism for frustration—the extra pelican chick, for instance?[23] Moreover, the total system of kenosis and remembering is in the end very close to that of process theology. The transcendence of God is still very hard to imagine. Do all theological roads in the end lead to something very close to process theology?

And do these roads also lead only to modifications of ends justifying means? Southgate hints at evil having its source in values we would not wish to relinquish;[24] this argument stops short of being a variant of "the ends justifying the means" only because he believes that animals are compensated for their lack of selving and transcendence. Here, too, is the danger Deane-Drummond has hinted at, that of normalizing evil in the development of theodicy, thereby accepting or depicting evil as necessary.[25]

Robert Russell and NIODA

Robert Russell, a physicist by training, has spent a lifetime examining the problem of how God, understood in biblical categories, might act in the world in a way that does not violate laws of nature. This, of course, lies behind the theodicy problem. If God can intervene but doesn't, then we worry about the morality of God. If God can't then this changes drastically the character of God. But if God can, and we are committed to some level of interaction then there still remains the problem of how, and whether the means of interaction sheds light on the problem of evil. For Russell quantum mechanics is a possible locus for what he calls Non Interventionist Objective Divine Action (NIODA). Quantum mechanics is essentially open, but it also becomes possible to imagine that in this mysterious place God might influence pervasively without in any way contravening "laws" of nature. This claim is controversial, though many critics also point to the fruitfulness of following this possibility to greater depth. Russell acknowledges that although this may potentially solve problems of God's intervention, it still does little for the problem of evil. There remains the hunch that if we were only to understand this quantum level better we might not only understand divine action but also its limits. One of those limits Russell suggests may be the decision of God not to intervene in the lives of sentient creatures, which of course returns us to the free will defense in some form with its attendant problems of God's influence on the non-human animal world. Moreover, because Russell traces the lineage of evil down into the world of matter, its tendency to decay, and to the second law of thermodynamics, he reaches a bedrock at which further theodicy thinking becomes impossible. At this point he appeals to the future, the impact of the future on the present, and the eschatological hope. [26]

One cannot, of course, disagree with the eschatological hope, but if that is the only place we end up we have very little which separates us from atheism. Atheists also wonder at the process of life on earth, but they see it as all self-perpetuating; it just is. It is neither good nor bad. With Russell I would acknowledge the beauty of the natural world and I would want to posit that we must understand, not creation necessarily, but God's present involvement in a way that allows us to see goodness as goodness, and not to be distracted in this seeing by the presence of tares, nor by one particular reading of the evolutionary story.

Sarah Coakley

Sarah Coakley has also recently entered into the area of theological dialogue with evolutionary biology.[27] Sacrifice, she argues, presently has very negative connotations, being associated with suicide bombers and alternatively within Christianity with the Girardian and feminist critiques of sacrifice, as well as Girard's insistence that there is a "primary violence deeply encoded in the roots of human nature," and that human social stability is founded upon this deep malignancy.[28] Coakley has been involved in a large Templeton funded project with Martin Novak, a Harvard mathematician. He has charted why there are mathematical reasons to expect that "sacrifice" and cooperation will enhance the fitness of a population, especially over a longer time period. Coakley makes the argument that what might be instinctual sacrificial behavior in lower animals is the biological basis for the possibility of altruism at higher levels. Thus cooperation or sacrifice can be seen as a third principle of evolution, at least a balance to the random mutation and survival of the fittest and its associated selfishness. Coakley persuades us that we are in this way given a new way of arguing for the rationality of Christian belief, because Christian values are congruent with sacrifice and cooperation, which are at the heart of the evolutionary process. As a result we also have a basis rooted in evolutionary life, she argues, for the case that sacrificial aspects of the atonement should be reintroduced from their recent feminist and Girardian-inspired oblivion. I would agree with Coakley that re-imagining evolutionary theory helps enormously in arguing rationally for Christianity. This is what I also intend to do in Chapters 7 and 8, using multiple lenses that all give a different view of the process than has popularly been described.

I agree also on the importance of highlighting the cooperative and sacrificial elements of the evolutionary process. These do make the task of reconciling theology with an evolutionary perspective a great deal easier. I would agree that instinctive patterns of maternal sacrifice, for instance, or other similar sacrificial stances may lay down the biological substrata that make fully conscious altruism and cooperation possible later in evolutionary development. But does this then put a sacrificial gloss over all the brutality of the evolutionary process? Where there is sacrifice there is also cruelty; the cruelty is only highlighted by the need for the sacrificial. Questions remain about the ruthlessness of a predatory system that makes a sacrifice of the most vulnerable members of a community. This might weed out the unfit but it also eliminates whatever positive characteristics these animals carry. There is an aspect of this work that comes close to normalizing the predatory because it is all seen through the perspective of the goodness of sacrifice, even though that victim was unwilling or unconscious and, in fact probably just wanted to live. Yet I would agree that symbiotic aspects of the evolutionary process are important, more so than purely sacrificial ones, because in cooperation that is fulfilling for all creatures or life forms involved there is a foretaste of the higher forms of divine and human communion.

Imperfection

John Haught believes that the creation is good, but imperfect, and on a journey toward perfection. "[S]acrifice can be interpreted in terms of the anticipatory, unfinished state of the universe," he says.²⁹ God recognizes the blind alleys of the evolutionary process and wishes they were not, but is ultimately—unlike process—overseeing a "final cause" of all creation, which will be the perfection of creation.³⁰ In a recent essay Haught looks again at the concept of perfection in Genesis, given that the old paradigm of perfection, fall, and redemption no longer works. He argues that the concepts of fall and original sin "remind us of the radicality of our need for redemption."³¹ He calls for the "transpositioning" of the concept of perfection from the past into the future, because the idealism the story engenders is still of immense importance.³² He points to the promise of the God of the future, and sees in the Genesis narrative hints of this forward bias. Ted Peters also speaks to this response to evil, offering a proleptic theology, which means that he understands our very being "as determined by, and defined by, our future."³³ This he calls "retroactive ontology." The emphasis is on openness to the future and to new possibilities, to seeing this as the upside of the contingency that otherwise wreaks havoc in our lives. He picks up that very dominant theme of the biblical text that speaks to the breaking in of the *basileia* (or kingdom) of God, something so alien and so complete and so loving that it does indeed seem to come from the future. Haught resonates with these same echoes in Genesis. To my mind, however, this is only half the story, and in being half the story it runs the risk of minimizing evil, or reconciling us to evil, as Deane-Drummond puts it.³⁴ We might ask what the world would be like if indeed we were moving toward perfection. Surely we might expect a universe in which overall progress is more visible. We might expect to see the point of suffering, the way in which it aids in the begetting of perfection. None of this is evident, though I would argue that we do see a growth in the wheat of life, often obscured by its accompanying tares. Kenosis, intratrinitarian kenosis, and imperfection, then, are true in part, but I would argue that they need to be augmented by an acknowledgment that the tares also are growing. Anything else gives us a very inadequate way of understanding the mix of perfection and corruption we experience around us and the perfection Scripture speaks of our being able to discern at all times (Rom 1:20).

Nor do these explanations show how it is that we can trust some aspects of the creation to reflect the presence of God and show evidence of the love of God when other aspects seem to be working so much against the flourishing of life.³⁵ The dark side of creation will be discussed more in Chapter 9, but here I note only that there is also in Scripture the persistence of a thread of tragedy; humans live with a notion of perfection and immortality always near yet always inevitably lost. And yet we are only participating more consciously in a dynamic that involves all of life and perhaps the whole of the creation.

Celia Deane-Drummond and Theodrama

I turn now to yet another recent attempt at evolutionary theodicy. Celia Deane-Drummond, like Southgate, makes a radical departure from the process theology of the twentieth century. Deane-Drummond draws very heavily on the work of both Hans Urs Von Balthasar and of Sergii Bulgakov, giving a new and fresh look at old problems. Deane-Drummond argues that previous attempts to look at God in evolutionary history have eclipsed Christology and have flattened out the drama of God's engagement with the world. God's active engagement and even battle with the darkness of the world is something she calls "theodrama" after the term used by Von Balthasar. If we think of process theology, for instance, this is certainly true. Christology becomes a variant of an actual occasion, a special case, rather than being that which constitutes the rest of the story. In dealing with theodicy Deane-Drummond comes to the point of saying that there is a *shadow sophia*.³⁶ Sophia (with capitals) is both the Wisdom incarnate in Christ and the wisdom of the creation, the intelligence working its way out in evolutionary and cosmic history. But there is a shadow side. Deane-Drummond does not develop this idea of the shadow, but it is the shadow that Christ overcomes and over which he has victory in the resurrection. For Deane-Drummond the answer to evil is ultimately in how we understand resurrection, redemption, and new life.

She argues for a way of approaching atonement which will give due credit to both objective and subjective aspects and to understandings which emphasize the love of God encompassing the whole of the created order, especially the animals. I would argue that we are very much in need of such a paradigm, and furthermore, that if we look closely at the Old Testament and the gospels this drama of Christ against the powers is evident, as it was in the early church. The idea of something being deeply amiss in the creation long before humanity emerges will be developed further in the next chapters.[37]

Functional Integrity and the Fully Gifted Universe

A related but more recent understanding of God, however, shares a lot in common with all the aforementioned understandings of God. It is close to deism and process in that it does not require or need an intervening God and God's full character will only be revealed in the future. Howard Van Till has argued that God has front-loaded, or "fully gifted" the universe with just the requirements for its ongoing life and flourishing.[38] In this sense God is omnipresent and all-powerful and loving, and God certainly anticipated what was needed but does not enter into the creation subsequently except in this pre-ordained way—although even here time itself lends an elusive character to what we might mean by *pre-ordained*. Now God's fully gifting the universe and therefore being present by God's Spirit is subtly different from deism. It must in some sense be true, and it is verified increasingly by the anthropic coincidences, the biology of constraints, and the ubiquitous emerging of new states of the universe.

The fully gifted universe is a way of saying that God is here, working all along, and at times new states or things occur that look to us as though God has intervened. If we had a God's eye view of reality though, we would see that new levels are emerging behind the scenes, working their way to appearance, much as we must imagine the lights of the universe emerging from the newly formed stars. What we once imagined as God intervening in powerful ways is really God with us by law. The universe itself is dynamic and fruitful. The fully gifted universe allows that God acts in ways that are always hidden in the mystery of laws and perhaps of laws not yet visible to us. And we must affirm that even the greatest and most dramatic acts of God in Scripture are visible and natural when we see them. The incarnation, for instance, is visible only as a human child and later a man whose story points beyond itself but in invisible ways. The fully gifted universe is also consistent with the wheat-and-tares universe for which I want to argue; for it affirms that whatever death or disaster the universe might seem to be heading toward there is always a gifting not yet visible within it.

The fully gifted universe, however, does not lessen too much the problem of evil, given an omnipotent and omniscient and good God; for we must question the gifting of such a universe. The idea of the universe's functional integrity and its front-end loading of giftedness is an idea that much of current thinking is grasping implicitly. It is an important idea, but still leaves us with a God who has thought of everything except how to minimize or mitigate horrendous evil over long periods of time. True, this idea removes the thought of an intervening God who doesn't intervene, but this God is really present all the time, not just in those moments of intervention that the popular mind might once have imagined. We might ask whether functional integrity does not also flatten out the sense of God's dynamic presence, removing the element of theodrama that Hans Urs Von Balthasar and Celia Deane-Drummond, after him, insist is a necessary part of the story of God.[39] But perhaps it is we humans, as God's agents, who should be presenting this element of the dynamic God in the world in our acts of dominion and healing of the world. These ideas will be taken up in Chapter 10.

Resurrection

Most of these theological approaches to evil end by including or emphasizing the resurrection and redemption of animals. This is true of Deane-Drummond, Southgate, Moltmann, Linzey, and many others. Most writers are somewhat coy about the extent and individuality of animal redemption, but many outline how an understanding that incorporates resurrection also changes Christian theology.[40] Resurrection first of all changes the emphasis, as Deane-Drummond argues, from atonement to redemption, for most categories of animals suffer but can barely be said to sin, even if their lives are characterized by the missing of the mark that defines sin at the most basic level. Deane-Drummond also points out that it is objective theories that are most helpful when it comes to animals. The redemption of animals does of course also imply a kind of universalism for humans; for if animals who cannot "appropriate" redemption are saved, so must people be who cannot or know not how to do the same. Such a change in the heart of Christian theology indicates how deeply Christian doctrine stands to be transformed when we take the nonhuman world into account.

Perhaps David Hart has said it best, that we must hate the devastation and the violence and the mercilessness of what nature often does. Without resurrection, without Easter, we cannot see God in any of this.

He says:

> ...the Christian should see two realities at once, one world (as it were) within another: one the world as we all know it, in all its beauty and terror...and the other world in its first and ultimate truth, not simply "nature" but "creation," *an endless sea of glory, radiant with the beauty of God in every part, innocent of all violence.*[41]

This is a claim that God is not behind everything in the world as it is now, but that ultimately God will redeem it all, and that through the eyes of faith the radiance of the world to come is even now visible. Hart writes in response to the Boxing Day tsunami of 2004. We are considering here another tsunami, the long history of violent animal death and predation and extinction. One part of the Christian answer is resurrection. But Hart himself, although a keen observer of nature, writes assuming the primacy of a human fall. How do we reorient Christian doctrine around a prehuman fallen world? How can we make coherent a world in which God promises redemption but that redemption is so long in coming, in which this long history of predation has been suddenly, as it were, brought into focus?

Conclusion

In this chapter I have outlined a number of theological responses to evil. Deism has become a default position with many scientists, as has process theology with feminists and eco-theologians. I discussed the consistency of these positions but their less than adequate tallying with Scripture and experience. More orthodox theologians included the theology of hope in the twentieth century, which affirmed that God has come to earth in solidarity with humans, and indeed with all life, to suffer with us. Others have extended the meaning of kenosis to emphasize the giving up of God's power that is required to make room for free creatures. Southgate places the kernel of kenosis in the very heart of God and in the need for the making of selves either in the Godhead or in the animal. Haught, Peters, and other theologians look to the future as the source of redemption and of the sense of perfection and idealism that is the beginning of the religious spirit. Celia Deane-Drummond opts for a shadow Sophia. For others it is the resurrection that is worth the long process of agony; theologians of animals have then begun to extend this notion of salvation into the animal world, calling on images from Isaiah, of the peaceable kingdom, and of Noah's covenant made with all flesh.

5

The Best of All Possible Worlds?

> *Help each other. Love everyone.*
> *Every leaf. Every ray of light.*
> *Forgive. Unless you love, your life will flash by.*
> MRS. O'BRIEN in *The Tree of Life*, director Terrence Malick

DAVID HART SAYS that God is not the author of the violence or the tragedy, although God is the one who will ultimately turn all evil to the purpose of good. At the heart of the conundrum we are discussing is the problem of whether states of affairs in evolutionary history are necessary. Did evolution have to be the way it has been? Was there no better dust, as John Cottingham has suggested?¹ An associated issue is whether we take evolutionary theory as complete with neo-Darwinism or we allow that other emerging "rules" might not modify evolution's overall philosophical thrust. Many theological responses to evolutionary carnage accept that neo-Darwinism is the best and complete description of how nature works and try to understand God's agency within these limits. Thus conservative approaches to the evolutionary/theology dialogue are more likely to affirm that this is the best of all possible worlds. Deane-Drummond's shadow sophia is an exception. Sarah Coakley's insistence upon the new rule of the sacrificial/cooperation within evolution is another. Here David Hart's voice is again relevant when he says:

> It is strange enough that the skeptic demands of Christians that they account for evil—physical and moral—in a way that draws a perfectly immediate connection between the will of God for his creatures and the conditions of earthly life; it is stranger still when Christians attempt to oblige.²

I will examine arguments for not accepting neo-Darwinism as the best and only way of looking at evolutionary history in Chapters 7 and 8. And indeed a wider understanding of evolution gives theology a lot more to work with—not just selfishness, but also cooperation; not just brute survival, but also creative adaptation, for instance. In the meantime, in this chapter I look at arguments that this is the best of all possible worlds, that the evolutionary progression really had to be the way it was, and that we should accept this with equanimity because to do otherwise is to question God.

Is the history of evolution from the beginning really a true representation of God's will in its entirety? If it is we are in a deeply puzzling place. For I see in creation, as Hart observes, a vast array of the troubling and the perfect. In fact before Darwin, with the fall as the explanation for all evil, nature was routinely plumbed as evidence of beauty and perfection. Today we know even more about that exquisite perfection. For this reason engineers have taken to copying the designs of creation in biomimicry.[3] The exquisite fine tuning that has brought us to the point of life and its flourishing is another evidence of the seeming forethought of a creator—multiple worlds notwithstanding. For this reason science is so exciting. There is always another remarkable adaptation to be discovered, deeper in to go, more puzzles to be unraveled.

Yet we are also aware of the other side, the darker side. In the time of, say Jonathan Edwards, one of the keenest theological observers of nature, the dark side of nature was explained away as a direct or indirect result of the fall. The collapse of this explanation has led to an attempt to incorporate all that we see that is not the direct result of human agency into the wider will of God. I suspect the attempt to baptize all of life's modes of being as God's will is wrong.

The Best of All Possible Worlds

Christopher Southgate argues that this is the life we live within and it pretty much has to be this way. Philosophers have long speculated that in some sense, imperfect though it is, this world we have may be the best of all possible worlds, even while they have urged against the naturalistic fallacy. In some small way we must say that this is not evident for any reason except that things are this way. Pelican chicks are sacrificed, but other birds manage to look after all their young, or they only produce one, as penguins do. Chimps are aggressive, but bonobos have found ways of limiting their aggressive traits. Southgate argues, however: do we really believe that God was thwarted in making a straw-eating lion?⁴ He does not want us to divide up the evolutionary process and its results into the good and the bad.

There is some point in this. It is hard to imagine a straw-eating lion. But then it is also hard to imagine the divine/human person of Jesus and life after death. I would be very cautious of pointing to this or that aspect of the past and marking it as evil or good—in the same sense that we are cautioned not to do this with respect to the wheat and the tares; the tares are inevitably holding up the wheat. Nevertheless, we can say that there are troubling aspects to the prehuman evolutionary history and troubling aspects to its mechanisms if God is even ultimately responsible. Some aspects of predation are concerning, especially those which involve systematic group stalking and the slow capture of an animal and the playful prolonged engagement with its prey that a cat displays. Nevertheless, living in this particular world as we do it is impossible to imagine how things might have been and what is necessary and what isn't.

In the novel *The Sparrow* by Mary Doria Russell, a novel that thinks deeply about these issues, there is an encounter on an alien planet between humans and two sentient species.⁵ One of these species has evolved to be aggressive. This is also the species that creates a material culture of great beauty. In *The Sparrow* the agrarian species is peaceful, has no technology, and is in turn eaten by the dominant species. Similarly, in our world the peaceful Amazonian tribe the Piraha some similar remote tribes make nothing at all or very little, and they are, relatively speaking, without violence, without jealousy, and without any desire for the material culture they see in others.⁶ Is there a correlation between the culture we value highly and a certain level of violence? This is a troubling question, one that has often been hinted at when philosophy claims that we may find that all suffering was worthwhile in the end for the making of value and beauty. Is the eating of the tree of the knowledge of good and evil what gives us all our beauty and all our destructiveness? *The Sparrow* seems to suggest that this is the case—that we pay a price for a high and advanced culture. Perhaps it could be otherwise, but only if we are prepared to surrender all that civilization and agriculture gives us. Perhaps this is engrained in the evolutionary process itself; every advance in complexity or sentience requiring a degree of knowledge of good and evil. Certainly some authors are prepared to argue that selfishness is required to form the creatures we are, but that this selfishness is neutral when it is only a part of the animal world.

Selfishness Is Original Sin

One recent writer who advocates this is the best of all possible worlds is Daryl Domning, a Roman Catholic palaeontologist who wrote with theologian Monika Hellwig. Domning claims to find no difficulty reconciling Christianity and his Darwinian view of the world. Domning and Hellwig agree that the traditional literal view of Genesis is no longer tenable. They see this, however, as liberation rather than a theological problem for evil. Domning explains that natural selection and mutation are the drivers of evolution and that this requires that all organisms act selfishly (as well as sometimes also cooperatively).[7] He thus rejects the idea that original sin is something rooted in the freedom of Adam and Eve, but he also rejects the idea that it is a cultural phenomenon. He worries that the "cultural transmission" view of original sin seems to invite the intervention of social programs and the possibility of sin's eradication by human means.[8] Domning locates "original sin" deep in the biology of humanity. He argues, however, that when this biological selfishness is operating at the animal level it is morally neutral, and the only way God could have made a world. When it operates in humans it is no longer neutral because humans can choose to act otherwise. He defends the suffering this entails in deep evolutionary time as being necessary to make the evolution engine work. All of evolutionary history is then blessed as good, not evil, because it all contributes to God's good creation. He says we should not apply our standards to the past. It is wrong in his view "to apply human ethical standards to the impersonal phenomena of nature."[9] Only when free humans choose selfishness is sin present. Nevertheless, Domning admits that the crossover between humanity and higher animals is blurred, and he is willing to accept that some animals do participate to some extent in moral categories.

Although I think there is a great deal that is odd in this argument, I think something like this argument is often assumed unthinkingly by Christians. We might ask though, why are actions morally neutral when performed by animals because God could not make them any differently, but morally culpable when performed by humans who often have very little awareness of what they are doing and the consequences of their actions? The argument of Domning and Hellwig only works if becoming human makes for such a radical shift in our consciousness and makeup that we are easily able to resist the urges that animals experience and cannot resist. There is very little evidence that this is the case. If God has made us this way and there is indeed no other way, then we are driven by deep biological traits over which we have little power, even if occasionally we might make a seemingly free decision. Human nature is distinct in its layers of consciousness and the ease with which we can construct layers of deception. Our deep motivations are often hidden to us. Because evolution posits that each level of consciousness gave way slowly to the next, God must bear responsibility for all of it, even if in some poignant shift in consciousness we come to share in the sense of shame and guilt for all we do. I do not doubt that the Edenic story does tell us this: that humans have crossed a threshold that takes us into the realm of bearing responsibility for our actions. But I believe, *contra* Domning, that if really this is the best of all possible worlds and God has decided that this was worth the cost, then humans, with animals, must bear a lot less responsibility than God.

Moreover, the grammar of original sin is undermined if it is linked to biological selfishness. Original sin stands for the idea of a rupture in human and animal history that extends the effects of a radically free decision to subsequent generations of human and animal. Necessary biological selfishness certainly has the same effect, but it can hardly be called original sin if it is far from being a rupture but is in fact a seamless part of the fabric of life. The grammar of original sin in the original Edenic discourse separates God from sin and evil; in Hellwig and Domning's view it implicates God deeply in the causes of evil and sin—but necessarily so. God could not have done any better.

Southgate and the Only Way

In a 2011 *Zygon* article Christopher Southgate expands on his own ideas about this being the only way that things could have occurred.[10] He interacts in particular with William Messer, who like me believes that the natural world as we know it includes something like a force opposed to God. Southgate very usefully refers to the difference between his own view—that value and disvalue are entwined but the latter is required for the former—and those of us who would ascribe disvalue to some alien force: "This is a key fault-line," he says, "in theology's response to Darwinism."[11] Nevertheless, interestingly, Southgate does refer to this state of affairs as "disvalue intrinsically and necessarily intertwined with value,"[12] a description that in many ways fits that of the wheat and the tares I will describe further in the next chapter. Southgate also describes the world of nature before humans as containing violence and suffering, both very close to being something more than disvalue, perhaps even sin.[13]

William Messer articulates this opposing force in terms of Barth's "nothingness." Like me, and unlike Southgate, Messer believes that there are elements in nature that are not part of God's purpose. "Das Nichtige," or "nothingness," he says, "is what God rejected and did not will, in creating everything that exists and pronouncing the creation "very good." As such, "nothingness" has a strange, paradoxical, negative kind of existence.[14] We all agree that there are aspects of creation that appear at least to be disvalues. We disagree on the provenance and the necessity of these disvalues. Southgate argues back against Messer that none of this really provides a theodicy, especially given Barth's reluctance to wade into anything like the evolutionary waters and apply this "nothingness" to the natural world. Messer goes on to say something very similar to the wheat and the tares when he adds, "the biblical witness, in short, requires us to say of the world we inhabit *both* that it is created *and* that it is fallen; *both* that it is the work of God, pronounced 'very good' *and* that it is badly astray from what God means it to be." My own reaction to the use of "nothingness" is that it is not personal enough. I have never thought that "nothingness," however subtly defined, gives sufficient weight to the horrors of the Holocaust happening so close to Barth's native Switzerland. Nor does it sum up the personal nature of temptation, as Jesus encountered it in the wilderness, at Golgotha, or in demons he exorcised. That all of this may in the end have no independent authority or life I can well grant. In the meantime "nothingness" is too empty a phrase for something so personally intent on death and destruction. I can, however, agree with Messer that, yes, something is opposed to God and is working in some way in the natural world as well, as the world of humanity. Southgate himself quotes Barth as saying "in the physical evil concealed behind the shadowy side of the created cosmos we have a form of the enemy and no less an offence against God than that which reveals man to be a sinner."[15]

Southgate charges both Messer and myself with dualism, which I will discuss at greater length in Chapter 9. Concerning Messer, he objects, in particular, to all application of "nothingness" or evil to creation—in spite of the creation being shot through with violence and suffering. Messer answers this charge by explaining that the doctrine of the incarnation means that God has joined us in suffering the shadow side of creation. I would agree with this, but would question whether that was a sufficient response.

But Southgate also responds to earlier versions of my own—appreciative—critique of *The Groaning*. He admits that he has relegated all talk of evil to that which is "freely chosen" by human agents. This is interesting because I would argue that much or even most human evil is not particularly freely chosen. Although human freedom means something, much of our lives are lived by habit, or the result of unconscious drives for good or bad, the result of personality or circumstances, choosing always to act as we will, but not to will as we choose. Even when we do choose we are often choosing in the dark, not really understanding or being able to see the consequences of what we do. Southgate does, however, admit that if something like human sentience emerges before humanity, then sin might also be present in the nonhuman world. I would argue that if the behavior of chimps is a natural progression of intelligence allied with predatory drives this then makes the predatory drives themselves not neutral. Then something like sin can be pushed further down into the evolutionary tree.

A Better World?

If I resist the idea that this is the best of all possible worlds, do I really imagine a world, a better possible world in which life evolves but without the carnage, without the predation, only the eating of plants, a world like C. S. Lewis's *Perelandra?* Do I really imagine that into this possible world a discordance, a brokenness, a rupture entered, not originally by human hand but by another? If there is a dark force behind physical reality or a shadow sophia, the question of its origins must arise.

There is of course some possible biblical backing for this in the difference between Gen 2 and Gen 9. In the original "very good" creation plants were given to animals to eat. This is intriguing, but it is too fleeting a mention to make of it much doctrine. In the end I believe there is something opposed to God, which must in some sense be God's creation, and that this something is beyond our story; it doesn't fit easily into any narrative that makes sense. It is too mundane to speak of prehuman "fall." Yet the opposition between this something and God is so great that redemption has a great deal to do with overcoming its effects, as Tom Wright insists.[16] This something, though, is not located in our world, but beyond it and beyond our story.

Whatever its effects there is a deeper level—in the second law of thermodynamics, in the tendency to decay that stands beneath this created ecological order and the new heavens and new earth. Even if there were no predation and all animals were vegetarian there would still be a creation groaning in need of redemption. This overcoming of the second law, as Bob Russell reminds us, and the hints of plant eating from Genesis, are just that, hints of new becoming.

Temptation

This opposition to God is most readily accessible to humans as the voice of temptation, often a tempting that comes in the guise of a friend—as did the original serpent. Whether the temptation is to National Socialism, to a whole new world order, to loyalty to a cult, or to excessive consumerism or self-starvation, the voice is ubiquitous and subtle. Whole systems and eras and cultures have been taken into its bondage. To answer my question then, I am convinced of the reality of this something and its subtle interplay with life at all levels, especially at higher levels of sentience. I believe this is why some people can look at the creation and see it as evil, as the main protagonist in the movie *Melancholia* does. But I don't think we can construct a mundane or useful chronology of this darkness, which becomes entangled with us, our lives, our capacities, and our fruitfulness at every level.

At some level, then, I do believe that this is the "only way" but not in the same way I *think* that Southgate does. Instead, I would argue that there is a fragility, as Michael Murray and others describe, to the ordering of the cosmos, a tragic element to the lives we lead, and this fragility makes this fallenness and the effects of this something all but inevitable. At the edge, at the "beginning," we get stories of the evil voice from outside us, from over the edge of our consciousness. The way I would put the situation—with C. S. Lewis—is that God has the "deeper magic" and always knows that this is the case, though the use of this magic is deeply sacrificial for God and for us—in Christ.[17] It is not that God knows the whole story and knows that the victim will be compensated. God knows that the creation can be made whole. Thus creation looks "cruciform," and creation looks as though disvalues are producing values, but really there is a modified dualistic struggle where the locus of the struggle is life and especially the hearts and minds of humans. With Celia Deane-Drummond, David Hart, and William Messer I would argue that the grammar of our theology must not accept what is opposed to God, and that versions of "this is the best of all possible worlds" do that. I would agree with Christopher Southgate, though, that this is indeed a fault line in current theology of nature.

Theology and Spirituality

However we disentangle questions of fallenness, violence, and culture, few of the theoretical responses satisfy at the level of spirituality. Perhaps they would be satisfying if all of us were equally afflicted in the interests of souls or freedom, or if everyone had only equal amounts of sadness in this so-called "best of all possible worlds."[18] Quite plainly we aren't equal. Great inequality of affliction abounds. Theology lives on in that space between philosophy and spirituality, attempting always to give responses that are sensitive to the individual soul. It heeds the arguments of the philosophers, but it is always pressing toward the spiritual and the need to satisfy that level of existence. Theology, however, has in some ways stricter criteria but also more resources than philosophy. Theology wants not only to acquit God but to be *convinced* of God's goodness and care. Thus Dawkins' arguments do not touch the philosopher but they do touch a nerve with theology. Piety demands that the hand of God be visible and discernible in some way; a possible state of affairs is not convincing.

But theology also brings further resources to the problem; Scripture quite openly discusses evil in the context of faith. In the New Testament the coming of the Christ is associated with a massacre of innocents, an evil so great that it is a portent of sorts of the worst that humans can do throughout history. At the same time it mirrors the countless slaughter of animals through the generations, animals who, like infants, were innocent (Matt 2:16). It is an evil that is not hidden and that causes one to ponder. Of all the events in history surely this, the incarnation, should have been the most protected by an omni-God. Clearly it was not. Mary made a hazardous journey hugely pregnant and innocent children were massacred because of the coming of the Lord. None of this is explained or justified, but its frank inclusion in the gospels is noteworthy. Faith was never proclaimed in a manner innocent or ignorant of horrendous evil. But neither do we ever get the sense from Scripture that this was the best of all possible situations, that infants *had* to die.

Nevertheless, in the coming of the Christ, in incarnation and in Spirit, God has come close to us and has born our suffering and taken this suffering into the very being of who God is. We do not suffer alone. There is a growing understanding that in taking on human flesh God was also drawing close to all flesh, for our bodies are studded with the signs of the animals out of which we have emerged. Read in this light it is significant that Jesus was born in a stable, with animals perhaps for company.

The persistent question remains: Why does God uphold the evil, which is in turn going to be redeemed? Why does God do it this way? These theological reactions to suffering only work if we can also understand the suffering of animals and the fierceness of nature red in tooth and claw. I would like to go back to the philosophers and note that another way of solving the problem of seemingly gratuitous evil is to reverse the argument and suggest that there is no gratuitous evil *because* a good God exists. To argue that a good God exists requires reasons. We must be able to *see* the purposes and hand of God as not entirely obscured behind the contingencies of history and evolutionary carnage.

Some of the theological responses give us reason to see the hand of God. Coakley argues for another way of seeing evolution; I also argue in this way. Deane-Drummond allows for a malignant aspect of reality, the shadow sophia; Moltmann, Haught and Hart expect us to see the perfection of the future in the present. Yet what we are presented with is a mixture of good and evil. I would argue that this mixture might be what we might expect in a "wheat and tares" universe. If we take seriously the wheat and the tares, we can trust our instincts about the good and not have them threatened or canceled out by the evil. This is the situation that has always been presented in Scripture—the coming of the Lord and the massacre of innocents—paradise and the ever-present serpent. While Dawkins might point to instances of evil and scoff, believers can respond that there has always been good mixed with evil in inexplicable ways and that in the end this evil will be overcome at every level of existence—for the lamb and the lion and the parasite and the tree and the ape and for us humans, as well.

Interestingly, this is also the theme of the 2011 movie, *The Tree of Life*, directed by Terrence Malick. It is an intensely postmodern film about the problem of evil. It begins with a quote from the book of Job. Its characters try hard to control nature in Waco Texas in the 1950s, but they do not succeed. Even the child asks, "Why should I be good? You let a child die." The message, however, is that there is glory and there is love among the carnage and evil. We have only to see it and to practice it. Hart and Malick agree that we can see the perfections of the universe as the hints of an essential glory that is being revealed. Moreover, humans appear to play a pivotal role in resisting evil on behalf of the universe. The tools and the weapons of the Spirit are the gift of Christ.

Conclusion

This chapter continues the theological examination of the problem of evil. It examines the fault line in theological responses to evil: is this the only way God could possibly have made a world? Is this the best of all possible worlds? The theological responses to evil are divided on this issue. Southgate in particular argues that our world is the best of all possible worlds. This chapter investigates his arguments and those of Domning and Hellwig, who believe that selfishness is morally neutral in the less than human world. In the end this chapter argues against the position that evil is necessary in a strong sense to achieve the kind of world we have today. This leaves open the possibility that good and evil have coexisted in the natural world long before humans emerged.

There is a variety of ways in which theologians come face to face with evil. Throughout all these responses the question persists: why did God make the world in quite this way, only to redeem it? For many there remains a persistent hunch that this is *not* the best of all possible worlds, even if that hunch is a possibly arrogant response of creature to creator. Nevertheless, like Job, we can believe in a good God in spite of all the evil if we can only see this God, and be convinced of God's goodness. This I argue requires an admission that the good and the evil coexist in history, long before the period of human becoming, in which all the goodness and evil have been enormously magnified and exacerbated. Such a way of seeing might give us confidence of the good God's existence because evil does not trump the good.

In the next chapter I examine the parable of the wheat and the tares and what it might mean that we live in a world in which tares are truly present.

6

The Wheat and the Tares

RE-IMAGINING NATURE

They say that God lives very high;
... But still I feel that His embrace
Slides down by thrills, through all things made,
Through sight and sound of every place;

ELIZABETH BARRETT BROWNING

IN THIS CHAPTER I introduce the parable of the wheat and the tares and argue that the parable can be expanded to include the whole of our cosmic existence, or at least the whole of the history of life on earth. In this parable of Jesus from Matthew 13 we get a picture of a world so divided and interconnected with good and evil that one can barely be distinguished from the other—yet the wheat and the tares are growing together. Although both wheat and tares are in the world, it is not easy to identify exactly which is which. If we take the parable seriously we should not even attempt to demarcate exactly or separate wheat from tare; we know, though, that not everything in the world or in nature is good. One approach to the ubiquitous mix of the beautiful and the tragic is to say that there is beauty and there is ugliness, there is perfection or goodness and there is corruption in creation and in life from the beginnings of life as we know it. This means there are always multiple ways of looking at the same things, seeing perfection, or seeing corruption. I would argue that theology needs to affirm the beauty and wisdom of God in creation. We must affirm what David Hart refers to as the "endless sea of glory," which is nature "radiant with the beauty of God" while acknowledging also the imperfections in nature as we experience it now.[1]

The argument of this chapter is that it makes sense to apply the biblical story of the wheat and tares to evolutionary history, and to the mix of good and evil we see around us and in nature. In this biblical parable, although humans interact with and cooperate with darkness they are not its cause or its beginning. Thus darkness in itself cannot be used as an argument against the existence of God if there are reasons to believe that a good and loving God does exist. Conversely if darkness is closely entwined with goodness, then the ambiguities of the evolutionary progression, although troubling, need not be ascribed directly to God. God can still be understood as working to redeem the suffering of animals as part of the larger creation. Thus the argument of Bishop and Perszyk that God can hardly be thought to be upholding the very evil that is then redeemed is overturned because God, is not responsible for the suffering of prehuman animals.

The natural world from the beginning has exhibited deep flaws, if not of structure, at least flaws in terms of integration. And yet nature also reveals extraordinarily dense levels of design and order and felicity and nomic regularity. I am intrigued, for example, by this statement from John Milbank:

> [T]o have a doctrine of the fall does not mean to believe that creation is totally corrupted or that it is corrupted in part. In a kind of negative version of the causality of gift, it rather remains entirely perfect—else it would not exist at all—and yet also entirely corrupted through and through.[2]

Nature is thoroughly perfect, and thoroughly corrupted. The eschatological thread of Scripture is that in the end these imperfections and corruptions at every level will be resolved in the marriage supper of the lamb, and the peaceable kingdom where the wolf and lamb will lie down together (Isa 11:6), and where swords will be beaten into ploughshares (Isa 2:4). If nature is understood as both the locus of the divine, as perfect, and yet also as corrupted by something other than God—yet not just humanity—then the death and destruction brought by nature need not cause us to despair of God. The mix we perceive means that the natural world is not an ambivalent irrational or opaque revelation of God; nature reveals God's perfections but it also reveals corruptions. Sometimes we can tell the difference and sometimes we can't. Human observers may disagree about the place and extent of animal suffering, but most agree that some first signs of the evil and grief and suffering we experience in human society began in the hominids that preceded us and are evident in our primate cousins, not to mention other animals.

To say that the world is corrupted is not to say that the earthquake and the tsunami and the eating of one animal by the other and natural disasters are necessarily corrupted aspects of nature in themselves. It is rather to say that there is not a perfect blending of the world being itself—with clashing tectonic plates and hurricanes and tsunamis, all interacting with humans and animals and other life. The delicate blending of physical constants and symbiosis that makes life possible at other times appears to betray us and all life.

Can we make some moral sense of evil by postulating a close entwining of good and evil, made more potent for good and ill when human sentience and intelligence became a part of the state of affairs? How does this understanding of fallenness survive when laid against the scientific story of origins and the long history of animal predation? In the next section I examine the parable and its interpreters and extend its insights to the natural world. I argue that this parable draws out three aspects of our experience and of revelation. First, it makes sense of the mix of good and evil we observe at all levels of the evolutionary progression. Second, the parable explains the close entwining of the two and the way in which goodness carries with it a dark side as often as not. Third, the parable reveals that the texture of good and evil, wheat and tares, cannot be seen without close observation. In Chapter 9 I examine the plausibility of postulating a deep entwinement of good and evil in nature, and I look at the wheat and the tares in terms of our discernment of God in creation.

The Wheat and the Tares—a Biblical Parable

Matthew 13 contains a number of parables about wheat and sowers and hiddenness. This is the account of the parable of the weeds [tares] from verses 24–30; 36–43:

> *He put before them another parable: "The kingdom of heaven may be compared to someone who sowed good seed in his field; but while everybody was asleep, an enemy came and sowed weeds among the wheat, and then went away. So when the plants came up and bore grain, then the weeds appeared as well. And the slaves of the householder came and said to him, 'Master, did you not sow good seed in your field? Where, then, did these weeds come from?' He answered, 'An enemy has done this'. The slaves said to him, 'Then do you want us to go and gather them?' But he replied, 'No; for in gathering the weeds you would uproot the wheat*

> along with them. *Let both of them grow together until the harvest; and at harvest time I will tell the reapers, collect the weeds first and bind them in bundles to be burned, but gather the wheat into my barn'."*
>
> *Then he left the crowds and went into the house. And his disciples approached him, saying, "Explain to us the parable of the weeds of the field." He answered, "The one who sows the good seed is the Son of Man; the field is the world, and the good seed are the children of the kingdom; the weeds are the children of the evil one, and the enemy who sowed them is the devil; the harvest is the end of the age, and the reapers are angels. Just as the weeds are collected and burned up with fire, so will it be at the end of the age. The Son of Man will send his angels, and they will collect out of his kingdom all causes of sin and all evildoers, and they will throw them into the furnace of fire, where there will be weeping and gnashing of teeth. Then the righteous will shine like the sun in the kingdom of their Father. Let anyone with ears listen!"*

This parable has obvious relevance to the church and to the *basileia* of the heavens (or Kingdom of God).³ Are we justified in interpreting this parable also within the wider context of the natural world from which it is, after all, derived?

Ecotheological Approaches

Recent years have seen the emergence of a new type of hermeneutic, reading the Bible through the eyes of the earth. Norm Habel has pioneered this approach with his *Earth Bible*[4], but many other theologians including Sallie McFague, Elizabeth Radford Ruther, Ruth Page, John Haught, Denis Edwards, Jürgen Moltmann, Holmes Rolston III, and others have also long emphasized the need to read from the perspective of the vulnerable earth.[5] Yet in the parable of the wheat and the tares, or the sower and the mustard seed, for instance, we have always jumped straight to the kingdom/*basileia* interpretation for not just human society but the church or people of God. The basic material of the parable is forgotten or overlooked.

Habel, in giving a new interpretation of Genesis, says:

> The *ruach* is hidden deep within the primordial domain, waiting in the darkness, with Earth... This story of creation embraces the presence of God as an integral part of the cosmos, perpetually moving within the cosmos.[6]

"There is no reason," Habel adds, "to assume this divine presence departs from...any subsequent domain."[7] Yet don't we often assume that because God is not "interacting" or not seen to be interacting, that God is not present? If God's presence embraces the cosmos there needs to be some other explanation for evil; God is not absent from creation. True, there is the odd passage in Scripture which suggests that God has something to do with the current state of affairs (Rom 8:20 for instance), if subjecting the creation to frustration refers to God, and removing access to the tree of life in Gen 3:22–24. Whatever these passages mean I assume that they refer to some restraining of evil in the present age. The gospels make clear that God is thoroughly at war with evil and that evil has no part in God.

Whence comes the long suffering of evolutionary history? Ecotheological interpreters argue that we have overlooked the common condition of life and its interrelatedness, the source of life in God, the importance of all living creatures to God's covenant, and the interconnections and mutuality that exist among all living creatures. They argue further that although the Scriptures are full of the evidence of God's concern for all life we have read the text in the past in an exclusively anthropocentric manner; this is the true anthropocentrism, not that of reading importance into fellow creatures, but that of ignoring their presence.

In a recent interpretation of the agrarian parables of Matthew, Elaine Wainwright speaks of "erasure and mastery" as characterizing our previous approaches to this and other Matthean parables, ignoring "ecological texture."[8] We have overlooked the dense texture of "place, time, and social location" from which so much can be learned. She quotes John Haught, who has suggested that the evolution of the cosmos is the most "fundamental mode of the unfolding of the divine mystery"[9] Wainwright and Haught both emphasize the earth as the locus of divine mystery and activity. Wainwright also points to the interaction of human and divine powers that bring about the abundance of nourishment on which we depend. She speaks of "multiple actants including hints of divine presence."[10] Also drawn out of this parable is the emphasis upon the end or the *kairos* moment of the harvest, not on the sapping of the nutrients in the present that the tares might effect. "Both the evocative and the ethical are significant elements of Jesus' preaching of the *basileia*,"[11] she says.

What can be made of this interpretation that is relevant to the parable in the context of evolutionary history? First, an ecological reading affirms that which I have been attempting to show throughout this book, that God's presence is potent in the creation; God's purposes are being worked out in the life forces and ecological balances and the mystery of the created order. A deist understanding of God's involvement is certainly not Scriptural. Second, I argue that these ecological readings suggest that in spite of the good will and good work of the human, alien degenerate seed or influence that mimics the good seed is ubiquitous. It enters into the place of great productivity and it is entwined with the good; Matthew claims that the bad seed was sown by the enemy, not by God. Third, therefore to see this seed requires the eyes of faith, but also patience because it will not be identified until the end. Thus, says Wainwright: "Rather than being reduced to univocal meanings related to human behavior, the parables are able to function in ways that draw readers/hearers into greater attentiveness to Earth processes."[12] They give us "whispers of divine presence with us/with Earth (Matt 1:23)."[13] This reading of Matthew's parables is consistent, therefore, with taking the subject matter of the parable seriously. When we do this, I would argue, the parable reveals a double message, the presence of God and the alien influence of an evil entwined with the good.[14] The mixed picture the world presents of perfections and darkness, perhaps increasing with sentient activity, is consistent with these multiple actants—God, the earth and its bringing forth, the animals and their participation, and through it all some harmful influence as well.

Jacques Maritain

Another wider interpretation of the parable is that of the French Jesuit philosopher/theologian Jacques Maritain. Writing in the early to mid-twentieth century, Maritain, whose historiography was detailed in his book, *On the Philosophy of History*, gives us a major contribution to the peculiarly Christian understanding of history, although history for him is the period in which humans have made a record.[15] He draws deeply from the parable in question. Maritain extends the parable to the whole of the movement of human history, seeing in it both the good increasing and the closely associated evil also increasing in intensity. There are, he says quite bluntly, "two internal movements in human history"—toward doom and the "Prince of this world" and toward the "kingdom of grace" that "follows the attraction of Christ."[16]

While acknowledging its relevance for both church and faithful, he also pushes for a wider and more universal meaning:

> But what I would like to emphasize particularly now is that the parable of the wheat and the cockle [weeds or tares] has a universal meaning and bearing which is valid for the world as well as for the kingdom of grace.[17]

Sometimes one period of history will appear to be overcome by the darkness or the tares; at other times only the motion of increase in power and goodness will be evident:

> And, of course, in certain periods of history what prevails and is predominant is the movement of degradation, in other periods it is the movement of progress. My point is that both exist at the same time, *to one degree or another*.[18]

While we might at some times be more or less aware of the presence of both forms of activity, both wheat and tares are always present and growing together. The influence of the good on the evil, and vice versa, is a universal feature of history and can be seen, he says, in that "great discoveries are usually paid for in human history by the reinforcement that a given truth receives from error preying upon it."[19] An example of this, he would say, was the mechanist philosophy associated with the good of sixteenth and seventeenth century science. This mechanical way of observing was liberating at first. We have since paid the price for this error as we struggle to free ourselves from its tyranny as an overall lens for reality. In this way we see a mix of the good and dark sides of revolutions and endeavors, as well as the associated hazards of utopia building, which often involves the attempt to separate out the wheat and is almost always associated with huge social and spiritual harm.[20] Maritain affirms: "at each moment human history offers to us two faces."[21] I would suggest that at any point of evolutionary history these two faces are also present.

Reinhold Niebuhr and the Wheat and the Tares

Much of Reinhold Niebuhr's work was written with the parable in mind. He would deftly weave competing strands in human and social life like pride and self-preservation, liberty and individualism, with a wisdom that does not make final judgments but recognizes the inside and outside of the same phenomenon, part wheat, part tare.[22] For Niebuhr, human repentance and God's ultimate judgment and resolution of history are the last words. Niebuhr directly addresses the wheat and the tares in *Justice and Mercy*. He argues that humans stand in the "river of time" but can also rise above it. This is the result of our peculiar and powerful consciousness. As humans we begin to think we are creatures who can create our own destinies. When this happens, says Niebuhr, "[the person] might forget how much of a creature he is when he begins to create."[23] Niebuhr then affirms that whatever our towering consciousness believes, we stand "under a sovereignty much greater than ours." Nevertheless, our own power of rising above can go on endlessly but is also liable to make us doubt the reason for existence. With a certain height, says Niebuhr, "if you get high enough…you know that the farm of the good farmer and of the bad farmer look equally like garden plots. All distinctions disappear."[24] This analogy is very apt. When evolution as a process is abstracted to a certain point we no longer look at it from the intermediate moral vantage point; distinctions are blurred. The parable should encourage us to look at life and history from multiple vantage points and to realize that sometimes we are blurring distinctions where they still exist. I will argue that this is the case when it comes to nature; we are blurring the good and evil by not distinguishing one from the other in the nonhuman world.

But for humans also it is essential, he argues, to make distinctions and not see everything the same. In the Old Testament we are told to "choose life," as though we always know which way is the right way. In the New Testament, he argues the parable of the wheat and the tares shows a much more ambiguous moral situation:

> The man sowed a field of wheat and the enemy sowed tares among the wheat. And the servants, following the impulse of each one of us, asked if they should root out the tares so that the wheat could grow.[25]

When they are told not to do this it seems as though "it violates every principle of agriculture or of morals. After all every farmer and every gardener makes ceaseless war against the tares…And we have to make ceaseless war against evil within ourselves and in our fellowmen."[26]

And yet this ambiguity is also true to human experience. If evil is rooted out it is often at the expense of the good as well. This calamity is observed every time a person is put in prison and denied freedom or a person is executed. Two lives are lost instead of one. The evil in the person might be stopped but so is the good. The danger the person poses might be ameliorated but the good the person will continue to do—in almost all cases—is lost to society and to their loved ones. Thus it is sometimes premature judgment that does the most harm, says Niebuhr:

> The suggestion is that a great deal of evil may come from the selfishness of men, but perhaps more evil may come from the premature judgments of men about themselves and each other...these wonderful words of Scripture suggest that while we have to judge, there is a judgment beyond our judgment, and there are fulfilments beyond our fulfilments.

He continues in the same vein:

> Consider how much more evil and good, creativity and selfishness, are mixed up in actual life than our moralists, whether they be Christian or secular, realize. How little we achieve charity because we do not recognize this fact.[27]

Niebuhr has not thought of evolution in all of this. His gaze is firmly on our social life together. The examples he gives are those of Winston Churchill whose early ambition was transformed into statesmanship and magnanimity.[28] Moreover, communism, he says is oblivious to the reality of human life and of this mix of wheat and tares. It attempts to pull up the tares of self-interest, not realizing that most of the good of society is also linked to this ambiguous state.[29] Thus Niebuhr's thesis is that we rise above in our consciousness, and in this rising above the river of time we judge, but this judging must be always modified by the realization of the mix of good and evil present in everything to do with human endeavor.

Again, although neither Maritain nor Niebuhr were thinking of nature, they have extended the parable from its original meaning, finding in it rich explanation for the deep ambiguities in our spiritual and social life. Similarly we can draw the parable in the other direction to cover the dense mix of the perfect and the troubling that is present in the evolutionary process and in animals' lives. Niebuhr's point that it is in seeing the details that the mix of good and evil emerges is particularly relevant. If we take too sweeping a view of nature or of evolution the pattern of good and evil can be missed. This is particularly the case when the concept of moral neutrality is applied unthinkingly to the whole of the natural sphere.

The Application of the Parable

Much has been written on this and other parables. The debate has centered around whether the parable applies to the world or the church or to both. Indeed where *is* the *basileia*/ kingdom, and who are the wheat; who are the tares or the weeds? Most exegesis revolves around these questions. Is it even legitimate to extend the parable to the natural world—from which, after all, the images spring? Some will deny that the Bible can be read this way, and indeed they will quibble with Niebuhr's and Maritain's applications to the social world and to history, and also with any ecological interpretation. But I would argue that it is in the spirit of holistic intertextual readings of Scripture to allow the extension beyond the first application to larger and more universal ones, especially if the subject matter is crying out for such a meaning. This appears to be consistent with the impact of Scripture as extending beyond the concerns of a small band of people to be much larger and more inclusive and to be relevant to all life and people on earth.

Thus in addition to the interpretation given by Jesus, there is much room for divergent readings. The three interpretations presented above give us some clues as to extension. The ecological interpretation urges us to consider the meaning of divine presence and yet also of something alien and harmful within the ecosphere itself, yet transcending it. Maritain and Niebuhr apply the parable beyond the kingdom or church to see it as a rule of social life. Together these interpretations can easily be applied to the evolutionary process. Thorns, thistles, and tares have "plagued" the natural world from the beginning. So also have discordant relationships, violence, and death. Only under human agricultural domination do thorns and thistles become an issue, and in human societies, especially agrarian ones, violence has been raised to new levels. The other side is also true; only under human cultivation does the beauty of a garden emerge. Other aspects of the natural world like parasites and predation and animal sickness have been tares that are more readily identified as independent of the human presence. Sometimes tares such as viruses that produce plagues can nevertheless subsequently become important aspects of an animal's genome, as explained in Chapter 8. Moreover, some transformed natural world is a part of the *basiliea*/kingdom to be—when the wolf will lie down with the lamb (Isa 11:6). There is a covenant with *all* of life.³⁰ Thus I propose that a reading of this parable that extends downward and outward is warranted.

And although none of the interpreters above generalize to the evolutionary process there is no reason why we should not do so and every reason why the parable applies, every bit as much as it does to the social sphere that both Maritain and Niebuhr articulate. They have given us an example of extending the wheat and the tares beyond the immediate concern of the parable. I argue that all of these parables are indeed deep enough that they can warrant such an extension, especially to the world of nature out of which come the images themselves. The parables are metaphors, and as such are extendable down and out. They are like kernels of truth awaiting our investigation and our response.

This has not been done before because our acute awareness of evolutionary pain and animal violence is very recent. All of a sudden we have been given a vista of nature that was easily dismissed in the past. Before Darwin, if such pain was recognized it was blamed on the Adamic fall. This pain and awareness of long periods of animal existence have led to the expansion of ideas of salvation from the human to the natural world as well. The importance of taking the earth into consideration in images of salvation has grown; the new heavens and the new earth have come into focus as being real and material, albeit a materiality of a transformed type. These images are fully consistent with glimpses from Scripture itself. Thus if the parable is truly describing the *basileia* of God, surely the overarching reference of reality, then the parable of the wheat and the tares can be extended to the whole ecological sphere and to nature itself. If the dark powers in Scripture are to be taken seriously, then so too must be their influence on nature, even if that influence or that bondage is enormously increased under human domination.

What if the "enemy," whoever that is, has sowed the tares already in the fabric of the living universe, or at least that part of it we know? Certainly humans have enormously exacerbated the differences between good and evil, but also the harm done in trying to separate the good from the evil, in rising above and thinking we are gods. This same terrible mix of good and evil does seem to characterize the natural world, as well as our social life, even if the mix is more subtle. Yet it is just as perilous to try to separate out the events and processes of evolutionary history and of the communities and animal species that have arisen within it, and to say this is good and this is evil. Certainly we cannot look at chimps and pronounce them evil because they have a kind of violence that most closely resembles ours, for chimps also have a drive and the curiosity that gives rise in humans to the cultural values we esteem. In the evolutionary processes there is a similar entwining of the apparent good with evil. Viruses, as I already mentioned, appear to have co-evolved with other life on earth. What is responsible in one age for an epidemic might become an essential embedded part of DNA in the future. The processes have produced life of enormous beauty and complexity, but they are also responsible for predatory drives and the extinction of the weak. It is possible to remain troubled by the mix of creativity and harm that jumping genes and retroviruses impose. We may be rightly worried if the very existence of life itself is the result of *only* these troublesome processes—death and predation and natural selection.

Reading Scripture in light of the parable other dimensions and glimpses of preexistent evil come to light—the serpent already brooding in the perfect garden, the Adversary who challenges God in Job, the ambiguity in Ecclesiastes and the affirmations that God's love is there in spite of the famine, or the fig tree never blooming (Habb 3:17). There is the promise in Romans that neither "death, nor life, nor angels, nor rulers, nor things present, nor things to come, nor powers, nor height, nor depth, nor anything else in all creation, will be able to separate us from the love of God in Christ Jesus our Lord" (Rom 8:39). Such statements assume that indeed there will be an ever present darkness that is competing with the love of God. Indeed temptation frames the story of Jesus' life—in the wilderness and at Gethsemane (Matt 4:1–6; 26:31–34). Jesus' birth, as I mentioned in Chapter 3, was accompanied by tragedy for the innocents. We are told that the rulers of this age did not know what they were doing in killing Jesus (1 Cor 8). We are warned about the principalities and powers (Eph 6:12).

Reading the parable of the wheat and tares in this way does not, of course, tell us the whole story. It doesn't tell us who or what is evil; the ontology of evil I believe will always be inscrutable. Images and metaphors and scenarios from the past—Satan or the Evil one or prehuman fall do not serve us well today, but the reality behind and beyond them persists. We find ourselves at the edge of the historical and the narrative, in the area of what Barth calls saga or myth.

A wheat and tares interpretation of unsettling aspects of the evolutionary process, and of animal life in particular, also give us a new way of looking at the good and the bad, so inextricably linked in nature. Southgate, for instance, points to the beauty and power of the cheetah and argues that predation has made this body, which we admire.[31] Surely that makes the process of predation good? A wheat and tares interpretation would say no; good and evil are intimately linked. If we imagine figuratively speaking, plucking out the tares the wheat would also be killed. Thus admiring the cheetah does not in any way make predation alright any more than admiring the soldier and his strength validates war.

The third aspect of Niebuhr's wheat and tares interpretation is the necessity of seeing the detail, so that good and evil are not clothed alike in the same garments. This is true of both the perfect and the more troubling aspects of animal life. Yet again, it is only in recent years that we have been inundated with disturbing as well as the inspiring detail of evolutionary life. Darwin glimpsed this, as seen in the passage about the *Ichneumonidae* at the beginning of chapter 3. Seeing only the imperfections, on the one hand, predation and parasitism and the sometimes clumsy looking assembly of the human brain or spine might lead us to think that the natural world and animal existence is just a kluge-like medley of ill-assorted parts.[32] On the other hand, if only the complex social and biological and evolutionary connections are studied, and all embedded in a universe that has seeded life, only the perfections might be noticed. Understanding from a Christian perspective requires that both be examined soberly. This is the mixed picture that David Hart addresses in *Doors of the Sea*, as described in Chapter 4. Hart links faith with being able to see the glory of the creation, which is now so ambiguous and at times so treacherous.[33]

Sensing the Beauty—the Marks of Perfection

Reading the parable of the wheat and tares within a larger and ecological paradigm allows us to return our attention to nature and to observe its perfections as signs and harbingers of God's presence, as are the deep structures of the chemical and physical make-up of created visible matter over which life emerges. We might see the ubiquity of the deep mathematical structure and the anthropic coincidences, or the forms and shapes of the geometry of life. The anticipation of cooperation and creativity at the heart of life, the vigor with which life preserves itself, all reveal to the eyes of faith a perfection that can be yet extended and perfected further under human dominion. The beauty of the wilderness can be transformed in part into gardens, the strength of the rock or marble is worked to make art and buildings. But the perfection and beauty are already there. This beauty and perfection are particularly evident among animals, and especially higher animals. Moreover each form of life has its own particular beauty and way of being.

The wheat and the tares allow us to recognize all this without having to ascribe to God directly the imperfections. Critics may point out, however, that some of what I might want to single out as evil is necessary for life. Some of the randomness of genetic mutation, or some of the struggle to the death in the animal world, seems to beget stronger, better models or eliminate sicker animals. Viruses kill and also contribute to life. All we know, however, is that this is one way these values have happened. They may not be the only way. In Chapter 8 I look at newer paradigms of evolution, which seem to place less emphasis on some of these more competitive aspects of the evolutionary process. We might expect to find the evolutionary process itself is a wheat and tares phenomenon.

If the story of Scripture and experience tells us anything it is surely that something dark and alien lingers in our midst. And even though we might disagree about the presence and extent of evil in the natural world nevertheless at some point the darkness does become evident. Some might argue that predation was intended by God. We might nevertheless ask, is the deliberate infliction of harm by chimpanzees something good in God's eyes when we know that all these urges toward aggression in our ancestors culminate in human life with its conflicting propensities?

Moreover, the emergence of aggression and violence appears to be unevenly distributed in the natural world. Chimpanzees are arguably more aggressive than some rare isolated hunter gatherer communities of humans like the Piraha.[34] This raises the question about whether the human propensity for evil emerged most potently in the settled agrarian life that was the beginning of the experiment in civilization 8,000 to 10,000 years or so ago. There appear to be hints of this as an epochal line in the story of Cain and Abel.[35] There are also negative suggestions of this link in the relative peace and lack of strong hierarchy in isolated hunter gatherer communities when compared with the hierarchies and violence of our highly civilized societies today. All of this is most consistent with continuity rather than with a strong point of fallenness between the animal and the human. It is consistent with the tares growing in some soils and in some contexts more potently than in others but always being present as the substratum.

Conclusion

In this chapter I have examined ways in which the biblical picture of the wheat and the tares can be extended to the whole of the created order, especially to life. I look first at three examples, two from theology and one from biblical studies, and argue that these applications can readily be extended down into nature itself. A wheat and tares view of world does not see all reality as kluge-like, a haphazard compromise assembled in an ad hoc manner over time. It does not gloss over the evil and see only the good, nor does it assume that all reality is in some sense the desire of God, the best of all possible worlds. I then begin the task of re-imagining nature and the evolutionary process and the lives of animals within this paradigm. In Chapter 9 I deal with the apparent problem of a Christian dualism that this image of the wheat and the tares can raise.

7

A Picture Held Us Captive

> *A picture held us captive.*
> WITTGENSTEIN

EVOLUTIONARY THEORY MUST be one of the greatest paradigm shifts humans have ever made, and in many cases still are making. The conceptual and moral and religious hurdles to evolutionary theory's rapprochement with Christian faith are significant and are still being negotiated as evolutionary theory itself is changing. In the early twenty-first century these hurdles are replete with words now so coded and loaded with ideological baggage that a coherent discussion is not easily had. Words like "creation," "design," "God," "purpose" are difficult to navigate, conjuring up as they do images of fundamentalism and of opposition to science. Arguments using these codes are easily dismissed as coming out of various positions such as intelligent design, on the one hand, or theistic evolution, on the other—positions that whole segments of the religious/political world have already accepted or dismissed as packages. Subtle arguments that emerged quickly after Darwin and even before are now much, much more difficult to espouse. We have, as Wittgenstein—and Jeremy Begbie—have described, been taken captive by a picture.[1] The picture is of a materialist, reductionist, randomly directed evolutionary process. This picture is embraced by atheistic Darwinism and strongly rejected—along with evolution—by creationists. One navigates in these waters between the Scylla of hyper-Darwinism and the Charybdis of purposeful theism. In this chapter I head into these waters, with some trepidation, convinced that the picture is changing in ways that are more conducive to theological insight.

In the previous chapters I have tried to show that knowledge of our evolutionary past and of the emotional and communal lives of animals increases enormously the toll of pain and suffering about which we are aware. But this knowledge also undermines the grammar of the most widely used theodicy, that of Adam and Eve and a sudden rupture with paradise in the fall of humankind. In Chapter 6, and again in Chapter 9 I argue that we must re-imagine the whole created order, especially life on this planet, as combining perfection and a shadow side, intimately related, even as the wheat and the tares are entwined. I argue that the Bible takes seriously the dark side of reality, even if the origins of darkness are obscure, and even if we cannot readily tell the wheat from the tares at any point in history. In spite of the darkness believers have found faith in God, not only because they have an explanation for all evil in fallenness, *but because they discern the hand of God in nature*. This is the religious discernment that has been undermined by Darwinism, especially atheistic Darwinism. The Central Dogma of Darwinism has given us a picture of a process that is moving on in a random and uncaring and non directional fashion in which the only link with the rest of nature is what survives and what does not. This is a picture of life almost at odds with the rest of creation. We have been taken captive by this picture. It undermines the sense of God's presence with us, because the most important part of our lives is life, and that, we are told, is essentially impersonal, or at best selfishly personal.[2]

Mary Midgley, however, in a recent book, and in an article in *New Scientist*, says: "Evolution has been the most glaring example of the thoughtless use of metaphor over the past 30 years with the selfish/war metaphors dominating and defining the landscape so completely it becomes hard to admit there are other ways of conceiving it."[3] More importantly for the subject of animals, this picture devalues the inhabitants of other niches on this planet. If we as humans have lost meaning, and a sense of the particularity of the space we occupy—perhaps the God-space—then we have little incentive to think kindly of the animals or to value their well being. What is impersonal and undirected has come to be, by a trick of fate, personal and driven human beings. A difficult and ambivalent relationship ensues with the rest of life, which is not yet at the point of sentience humans have achieved. As Midgley says, we often relegate animals to the impersonal process and ourselves to some exalted state above them.[4] Simon Conway Morris has said of this picture, that we have been "in the last century trying to square the circle of a meaningless process, that is, evolution, leading to the appearance of a sentient species that sees meaning all around itself."[5]

For believers taken captive by this picture there is dissonance and tension as one attempts to harmonize a Scripture that speaks of God's care with the dominant evolutionary metaphor of selfishness and competition.

Without a discernment of God's existence and care comes a barrenness and lack of depth to our lives but also to animals and to our relationship with animals. While Genesis may have justified the worst offences against animals, a proper understanding of dominion and of God's intentions and presence in creation, as mentioned in Chapter 2, lies at the heart of a new perspective. Yet we can only be confident of God's care and presence if there is some way of "explaining" the presence of evil, or at least adjudicating whether evil comes from God, from some evil power or from human action. I have argued in previous chapters that neither God nor humanity is primarily responsible for most of evolutionary evil, and the attendant suffering of animals within it.

In this chapter and the following one I try to show that if viewed in a new perspective evolutionary theory is consistent with seeing "the embrace of God" in nature, as Elizabeth Barrett Browning suggests. Acknowledging the embrace of God, however, requires that we break the captivity of the picture that has settled thus far over civilized and ecclesial life.

At stake, therefore, is confidence in our ability to sense God in nature, a major reason for believing in God now and in the past. The *divinitatis sensum* can neither be argued for, nor negated empirically. Calvin argued directly for such a sense when he says, "There is within the human mind, and indeed by natural instinct, an awareness of divinity."[6] This is, however, not the full story, for although *divinitatis sensum* can be considered a part of our human make-up, it may, like other human capacities, be alternatively undermined or encouraged by the appropriate social discourse, and is initiated or not, in large part by discernment of God's beauty and power in nature. If God is present with us in nature, a great deal of evil can be endured. If God is not discernible in nature all other affirmations of God's existence are easily relativized.

A large part of discerning God in nature is in our relationship to the animal world, though it is also to be found in the beauty of the landscape and in wonder at the way both life and matter fit together.[7] Thus the alienation many feel from nature is in part the result of this thin description of reality that neo-Darwinism can bring about in spite of the superbly written books by the likes of Richard Dawkins, Stephen J. Gould and David Attenborough. (The New Atheists are quite willing to admit and to acknowledge the numinous but this is not equated with any known or previous understanding of God. It is *their* numinosity.[8]) This alienation has helped us to treat animals as though they, like us, were just machines, to lose the sense of mystery that is associated with all living. Reenchantment of the natural world, re-imagining the evolutionary process can also draw us back into a stronger sense of kinship with the animal world. It is this kinship and wonder at the way an animal is in itself that is behind God's final statement in Job. It is the way an animal is in itself, flourishing and alive, that Christopher Southgate has described so well in his concept of "selving."[9]

Job and Theodicy

In the most important theodicy in Scripture, the book of Job, God says to Job in chapter 38:

> *Where were you when I laid the foundation of the earth? Tell me, if you have understanding. Who determined its measurements—surely you know! Or who stretched the line upon it? On what were its bases sunk, or who laid its cornerstone when the morning stars sang together and all the heavenly beings shouted for joy?...*
>
> *Have you commanded the morning since your days began, and caused the dawn to know its place, so that it might take hold of the skirts of the earth, and the wicked be shaken out of it?...*
>
> *Have you entered into the springs of the sea, or walked in the recesses of the deep? Have the gates of death been revealed to you, or have you seen the gates of deep darkness? Have you comprehended the expanse of the earth? Declare, if you know all this.*
>
> *Where is the way to the dwelling of light, and where is the place of darkness, that you may take it to its territory and that you may discern the paths to its home? Surely you know, for you were born then, and the number of your days is great!*
>
> *Have you entered the storehouses of the snow, or have you seen the storehouses of the hail, which I have reserved for the time of trouble, for the day of battle and war? What is the way to the place where the light is distributed, or where the east wind is scattered upon the earth? Can you bind the chains of the Pleiades, or loose the cords of Orion? Can you lead forth the Mazzaroth in their season, or can you guide the Bear with its children? Do you know the ordinances of the heavens? Can you establish their rule on the earth? Can you lift up your voice to the clouds, so that a flood of waters may cover you?*

God is saying that Job should understand that even though he has suffered deeply he must, nevertheless, believe in God, and indeed God's justice and goodness, *because of what he sees in nature*. Who else could have made all these beautiful and powerful and poignant creatures? God is saying that Job must trust because there is so much evidence of God's goodness and power and because he, Job, is so lowly compared with this God. Why should Job even think for a moment that he could plumb the depths of evil? The evil has no possible human answer, but the goodness and power is there. The Bible is quite frank about the extent of evil and its scope. But in the past faith has had a last say as it surveys the beauty and power and liveliness of the natural word.[10]

This way out of the theodicy problem is not accessible now without effort, and is barely compatible with the present metaphors associated with evolution. Darwin has given us an alternative answer for origins, for the majesty that Job encountered. This random competitive process does, of course, give us a kind of answer to the problem of evil in proposing the absence of a divine being, but does not answer why it is that our inmost moral sense is so opposite to that which is often seen to propel the evolutionary process onward. Darwinism has given us a view of nature, not filled with magnificent creatures but with randomly produced predators and victims of the evolutionary race. This touches a nerve because it is in the area of nature that so much is at stake both for believers and for atheists.[11] So long as the evolutionary process really is obscure and to all intents and purposes completely random and directionless and purposeless the atheist does have the upper hand, and may well be quite content for believers to affirm whatever they like in their own private space, if that is what they are so deluded as to want. But woe betide the person who argues that nature itself speaks of the glory of God.

Yet seeing things in quite the dreary way just described requires that hyper–Darwinism is still intact, and that random mutation plus selection and intense competition for survival is the extent of what is going on—as well as a really clever algorithmic way of reproducing life. If it is possible to see the evolutionary process in a substantially different or augmented way the old argument from Job can be more accessible, especially if we have come to believe, as I have argued in Chapter 6, that goodness and evil are intertwined at all levels of existence. Belief in God is never compelled by the evidence, but it can be consistent with it, or even likely and therefore rational. I would claim on this basis that newer understandings of, and metaphors for, evolutionary theory, when taken together and in summation have profound theological significance. They break our captivity to a naturalistic mechanistic competitive picture of nature. Whether or not they amount to a biological change of paradigm, for theology they do amount to such. These newer evolutionary dynamics—discussed in the following chapter—can make faith more coherent at the intellectual level, making it possible for us to trust our deep instincts that the beauty of the world is the work of a creator.

The Hiddenness of God

> *The characteristic invariant forms of systems like this are commonly "hidden"—they are mathematical rather than visual... sand-pile avalanches... the apparently random sizes of slope collapses are governed by the "hidden regularity" characteristic of self-organized criticality, distinguished by power-law statistics.*[12]

God, however, will never be transparently present in the natural world. The Trinity allows us to think of the natural world as God-breathed, yet with many of its dimensions hidden to us; the Trinity is part ground of all being or totally other, part incarnate wisdom part indwelling Spirit; and all one God. The biblical and theological traditions speak of a God who is partly revealed and partly hidden. In the Old Testament prophecies of Christ, for instance, while Christians can see the whole of the law and the prophets as pointing to Christ, the revelation of the messiah was so obscured in this revelation that the actual coming into history of the redeemer was surprising and scandalous to almost everyone. While Jesus gives signs and miracles of his divine/human status, these are given, we are told, to a largely uncomprehending multitude. Søren Kierkegaard reminds us that the problem of seeing God in our midst is not greater for us now than it was for the contemporary of Jesus, because the revelation, although decisive, was always veiled.[13] Not that his revelation was completely obscured or hidden. Jesus was not a man indiscernible in *any* way from any other man. Interesting also, that the New Testament mixes the mythical—prologue to John—with the ordinary and the historical; the coming of divinity amongst us, although observed as human, was, nevertheless, only describable across this spectrum of discourse.

"Truly you are a God who hides yourself," says Isaiah (45:15). "Where were you when I laid the foundations of the deep," says God in Job. "But God chose the foolish things of the world to shame the wise. God chose the weak things of the world to shame the strong" (1 Cor 1:27). Christ, believed to be the Son of God, came as a man, God hidden amongst us. "They knew him not" (John 1:10). What we attempt to nail down in biological or theological terms is ultimately not objectifiable and can be known only obscurely and peripherally—*Deus Absconditus*. This is the God of Luther who can be known dialectically, beneath the indignity of the cross and shame. Brian Gerrish discusses the two senses of hiddenness attributed to Luther, that within the revelation, and that beyond or behind revelation.[14] God hidden within the natural world is related to but not identical to hiddenness in the first sense, which Luther took especially to relate to the paradoxical hiddenness of God under and within foolishness, and behind the cross. Gerrish argues that Luther was aware of "an awesome, creative power quite other than the God [which] he encountered in Jesus Christ." This God was both transcendent and present in *each kernel of grain*.[15] The way in which God could be present in a paradoxical sense allows a depth to the revelation of God. God's presence is not always to be straightforward. It might be seen only through the eyes of faith, a discernment made about the true nature of things.

Schleiermacher also argues that God's activity is present, though hidden, "a power which expresses itself…at particular points according to laws which, if hidden from us, are nevertheless of divine arrangement."[16] Hiddenness is also paramount in the revealing and unrevealing God of Barth:

> We thus understand the assertion of the hiddenness of God as the confession of the truth and effectiveness of the sentence of judgment which in the revelation of God in Jesus Christ is pronounced upon man and therefore also upon his viewing and conceiving, dispossessing him of his own possibility of realising the knowledge of the God who encounters him…[17]

God can only be known on God's terms, and not in a human frame of reference. God's spirit may be with us, and within the natural world, but God is never objectifiable in the way that nature appears to be. The heavens declare the glory of God, *and* they obscure God's real character. Like a beautiful tune played by indifferent musicians it is sometimes possible to hear the perfections, and at other times the dissonance predominates; and the music and its beauty is not heard at all.

John Haught attempts to explain why the universe is so subtle, why God is so hidden within it by appealing to the unfinished nature of the present totality of things. "The unavailability or hiddenness of God" he says, "is in some sense, I think, a function of the fact that the universe we live in is still coming into being."[18] Although God is hidden both Berkhof and Elizabeth Barrett Browning and others expect that God is not *completely* hidden. The order is a reflection at an intuitive level of patterns and symmetries and beauty that are suggestive of mind, and perhaps even of love. Certainly Jonathan Edwards, the great American theologian of the eighteenth century understood the symmetries of the natural world to be a lower form of love.[19] It may be at this level that the intuitive connection to nature takes place, and is most important in validating the *divinitatis sensum*. The perhaps always necessarily intuitive connection is also always both hidden and revealed, able to be affirmed, but also to be denied. The hints and suggestions of God's intelligence in nature would be within the continuum, not necessarily the result of sudden breaks in or into the natural world in a process that otherwise continues. To see God partly hidden in nature is to have disclosed by faith a part of the character of this nature that might otherwise be obscured, not to see in it breaks in a materialistic process.

McGilchrist cast some neurological/psychological light on the way in which much that is true is nevertheless hidden. His thesis in his recent book, *The Master and his Emissary*, is that our civilization has been taken over by a left brain way of thinking of viewing. The left brain is the expert in analytical abstract thinking that is divorced from context and is mechanical. The right brain is more attuned to life, and to context and to emotional expression, and to first impact. The left brain, however, has a tendency to take over the thinking process, not to allow the right brain perspective. The right brain, in contrast, does involve the other hemisphere in its deliberations. Thus the right brain is the seat of our understanding of deep intuitive emotionally related aspects of reality, and the fact that we often do not have eyes to see or ears to hear is not surprising at all.[20]

Natural Selection

I have argued that we must discern the presence of God *in* nature, by faith, but not fully or comprehensively nor in an unambiguous way. Does this make sense at all in light of our present biological and palaeontological disciplines? Although long preceding Darwin in their incipient forms, deistic understandings of God, design options and also process theology have all flourished partly in response to the perceived repercussions of mutation and natural selection. In Deism God can be understood to have set the process in motion; natural selection is the ultimate deistic tool. Mutation and natural selection does not have to be consistent with any divine character because God is not directly involved, and has presided over a more or less mechanical law-like process. Intelligent Design opposes the implicit randomness and blindness in natural selection, and has grown partly in direct response and opposition to the metaphysical conclusions of Dawkins and others. But it too proposes mechanical like interactions by a divine hand in the evolution of life. Process is a theology that is gently consistent with the idea that God might be a weak indiscernible lure along with pure chance at the quantum or genetic level. Thus theology has attempted to accommodate or refute the most random interpretations of adaptation, though in doing so has developed theologies that are not ideal in terms of scriptural integrity or faithfulness.

Random mutation and natural selection is a slippery creature. Random mutations and best fit can be made to sound very impersonal, independent and autonomous, selfish and intent upon survival only—much like the ideal human of modernity. But random mutations of what? Is there anything else involved? Recent discoveries would imply that evolution is more complicated than this, and in the modifications and qualifications lie a hornets' nest, or at least the possibility of a paradigm change. That the whole process is less impersonal and less random is suggested by evolutionary convergence and emergence, and also by the physical and chemical laws of the matter that forms life; these will be discussed further later in the chapter.

At a micro level there is evidence for environmental feedback to the gene, for some genes being more evolvable than others, and for environmental influence on gene expression.[21] When the chemistry and physics of life are examined there are seen to be significant extrinsic constraints on the evolutionary process—even in chaotic forms—such as would reveal a clear direction. Perhaps an image might be helpful here. A coin spins down a museum funnel quite randomly but will inevitably end up in the centre. A coin is tossed into the sea and lost. Both are random but one is constrained in such a way that the result is far from accidental. Of course a funnel is not the best image. We might imagine a landscape with multiple funnels or holes, the equivalent of our niches or spaces. What is equally important with the random walk and the best fit, however, is the obscured existence of these niches, together with mathematics and communication/information. Convergence and other newer evolutionary emphases suggest the presence of a metaphysical funnel, perhaps largely hidden in the mysteries of the gene, or its development, or its placement within the larger whole—a random but constrained process. Suppose, says Conway

Morris, that "evolution is effectively the motor whereby the deeper realities of the universe may be uncovered..."[22]

Suppose indeed. This raises the question, should we be attempting to reconcile faith with the most random and impersonal interpretation of evolutionary mechanisms when the hidden niches may be just as important metaphysically? New emphases in evolutionary theory may hold the clues to recovery, or at least to tentative proposals that allow more theological consistency. As Jan Sapp has said of evolution:

> Revolutions are rarely complete, and the Darwinian revolution was no exception. Darwin convinced many biologists of evolution, but he did not convince them that natural selection accounted for the origin of species.[23]

Richard Dawkins and John Haught's Response

Into this vexed situation someone like Richard Dawkins claims to give a very exhaustive account of how organisms, and in particular the human being have evolved. He is absolutely sure that this is the unguided unpurposeful and flat process of random change and survival. Yet he himself is somewhat in awe of evolution, almost reifying it to the level of Evolution so that we could quibble over what it is that Richard Dawkins really believes. More than this, of course, Dawkins believes that faith is dangerous. Now in response to these and other claims of the New Atheists,[24] John Haught has responded that we must go "deeper than Darwin," that what the Darwinists say is all very well but there is a deeper level of mystery beneath and beyond what we see. Haught says for instance, "a vein of permanence abides behind the veil of all cosmic becoming and loss."[25] He says, "I want to reflect on evolutionary science with a mind and heart molded by a sense that nature is seeded deep in its fertile subsoil with limitless possibility."[26]

I would agree, and I would even suppose that this is the reason why Dawkins and other atheists are left in a state of awe. But if Dawkins were to ask believers then why can't we identify the God undergirdedness of the process, Haught would respond that it is because of God's "letting us be," letting the process be, letting life get on with being itself. Haught is content to affirm the existence of the deeper domain, while at the same time being quite adamant that we don't need to see it. Dawkins also mockingly asks the Christian why there is so much wrong or less than optimal in the evolutionary product if God is present but deeper than we can sense. Believers have responded that things are the way they are because of organisms being free and especially humans being free. Haught also adds that there is suffering because the world is unfinished and therefore imperfect.[27] Speaking of another atheist he says:

> And since neo-Darwinism has shown that design in organisms can be economically explained by natural selection working on minute random genetic changes over long periods of time, there is no longer any role for a divine designer to play. Therefore, it seems to Crews, evolutionary biology has exposed once and for all the utter godlessness of the cosmos... he shows no awareness of the power of layered explanation or explanatory pluralism, even though science itself employs it often.[28]

In the end the Christian response can begin to sound like special pleading. We can't see God and we don't need God—they say—as an explanation for all there is, and even though we excuse God when God does not respond to prayer or fails to give protection, even though the good suffer and the evil prosper, we will still believe in God—in spite of the acknowledged hiddenness of God. I think we have to say more than this, otherwise we will continue to be held captive by a picture. I agree with Haught's "deeper than Darwin" argument. This is why I think we need to challenge the atheist in the area of discernment. It is never enough to say that there is a lot that is deeper than Darwin going on as we move toward perfection, but don't expect to see any of it; don't expect to challenge the dominant metaphors.

Peter Berger has described the reformation as the time when the world become disenchanted; no more holy water, no more sacraments, no more nature as sacrament, only the Word of God, and that vulnerable.[29] Yet even in the eighteenth century Jonathan Edwards was fully confident that "images and shadows of divine things" were discernible in nature. This is because although he was reformed and post reformation he was also a mystic at heart, a scientifically and mathematically trained mystic with keen powers of observation, and he had a foolproof explanation in the Adamic fall for that which was *not* right in nature and in society and in humanity. Only now, post reformation, post Darwin do we have this puzzle about nature. It no longer unequivocally speaks to us of the glory of God. Hans Urs Von Balthasar has said of us: "Modern Man has had the frightful misfortune that God in nature has died for him."[30] Haught, similarly says: "A kind of cosmic despair has increasingly wormed its way into the high expressions of western culture." So we think, yes it is beautiful, but we know that deep down it is the product of random processes, and we don't have an explanation any more for all that is wrong. Kenosis and "letting be" is a lot harder to understand than was the Adamic fall. The sense that God is undergirding and upholding the creation is often weak.

I believe that unless we do better at pointing to the evidence of these deeper processes—hidden though their meaning is—we will have lost the metaphysical battle for the "picture," and we will never fully appreciate our intended purposes and connection with the animals and the rest of life.

The Deeper Processes

What then do I mean by pointing to the evidence of these deeper processes? It is possible to isolate out a number of aspects to this intuition. These will be described in more detail in the next chapter. The first is the deeply mathematical nature of all of life, including evolution. The second is the non reducibility of it all—the existence of the depth that is now described in terms of emergence and interconnectedness. The third aspect is the discernment of purpose or direction even if the nature of that purpose is obscured. A fourth is the coinherence and cooperation and symbiosis that so characterize all of life and that have been so eclipsed by the picture of the selfish gene in the last three decades. All of these lenses allow us to break the sense of being held captive by the picture we have hitherto embraced, allowing an awareness of depth as is possible in the stereoscopic images that can emerge from a seemingly random background. If the captivity is broken then the voice of God and the wonder of the creatures can be heeded by us—as they were by Job.

Whatever the mathematical substratum, however, it is very certainly not evidence that evolution didn't happen; it is not evidence that mutation and natural selection isn't at play, but it is evidence that life, mathematics, physics and chemistry are all deeply related. Life is not at odds with the underlying physics and math of the nonliving world. I would agree with Haught that the last thing we want is the specter of a huge machine universe. The mechanistic universe was the dark side Jacques Maritain saw in the rise of science. John Brooke has shown us how in the early modern era people marveled at the machine universe and saw it as a sign of God, but gradually the machine took over and we do not see it in that way now.[31] As a glimpse or a shadow of intelligence, this layer of order does work in a way similar to mathematics, which is after all a discipline itself rooted in paradox and difficulties at the level of infinity. To say that this difficult discipline undergirds all life is not to say anything easy.

In this next chapter I will proceed on a journey through some new insights in evolutionary theory, insights that I will argue have profound theological consequences. The first inkling I had that evolutionary theorists were thinking against the grain of the Central Dogma[32] was in the late 1990s when I heard a talk by the American Indian paleontologist and ecologist Dawn Adrian. She was arguing that postmodern notions of evolutionary theory encompass an evolutionary feedback loop into the genome.[33] Even at this level I could see that this broke the hardness of the Central Dogma a little. Anything that made the process less of an automatic unfolding against the odds, in which the environment only mattered by the passive dying or surviving of organisms within a fitness space, was helpful theologically.

I had grown up with the work of Teilhard de Chardin, but the more recent works of Philip Clayton and, later, Jacob Klapwijk give much more credence and detail to the intuition that the natural world exists at different irreducible levels, all of them interconnected, but participating in their own idionomic laws (laws appropriate to a given state), as Klapwijk describes it.[34] The work of Simon Conway Morris has also been of enormous importance in identifying the forward directedness, and the overall dimension of purposefulness that can be seen in the evolutionary process. And the ideas of cooperation and symbiosis are now leaping to the foreground of evolutionary thinking from many directions. Sarah Coakley, as mentioned in Chapter 4, has made these metaphors of cooperation and sacrifice central to her understanding of evolution and its reconciliation with theology. Jan Sapp has reminded us in his history of evolutionary theory that these metaphors of symbiosis have always been around.

Conclusion

I argue in this chapter that we must be converted from a picture of ourselves and of the evolutionary process that is almost necessarily atheistic. This picture has, in the words of Wittgenstein, held us captive. The loosening of this picture does not mean that God is any less hidden, any less other; rather it means that we have the option of reimagining our history, our evolutionary past, in a way that allows the rationality of belief, does not contradict our deepest religious impulses, and reinforces the sense of connection to and responsibility for the rest of creation.

8

New Dynamics in Evolutionary Theory

> *Evolution may be a fact... but it is in need of continuous interpretation.*
>
> SIMON CONWAY MORRIS

ALTHOUGH THERE HAS always been deep discussion about what evolution means, the narrow metaphors available hitherto make surprising the heterodoxy of evolutionary theory since Darwin. The philosophical divergence is evident even in the difference between the Dawkins of this world and the position of S. J. Gould. Gould sees the process of evolution as essentially contingent, and with punctuated bursts of activity, and is thus in awe of its continuity over three billion years.[1] Dawkins emphasizes slow change and necessity and—with Conway Morris—would be sure that were we to turn back the clock 3 billion years we would get essentially the same pattern of development and of species that we have now. These considerable differences between experts are interesting.

In *Genesis: the Evolution of Biology*, Jan Sapp unmasks the history of dissenting opinion in evolutionary interpretation in the last one hundred and fifty years and its sometime links to reigning ideologies, giving us a window into a world of controversy and disagreement at a philosophical and even theoretical level amongst biologists who all, nevertheless, accept the tree of life, common descent, and evolution by incremental change. Sapp also exposes all the threads and questions that have long prevailed and are still open today, as well as questions that are particularly highlighted in the present stage of knowledge. He explains: "Some nineteenth century evolutionists espoused several mechanisms and some included an unknown force through which evolutionary changes would be directed along a straight line."[2]

The persistence of Lamarckianism, for instance, is surprising. It has had a strange parallel life in France where it was popular because it was more compatible with religious and teleological views.[3] The Marxist regime in Russia advocated it fanatically and rejected all genetic theories for ideological reasons.[4] Yet in recent years it has been acknowledged that there is still some validity to Lamarckian ideas broadly understood. A new field of epigenetics has arisen in which environmental influences can change gene expression for subsequent generations. While the genetic code itself is not changed, gene expression is, because the environment or gender is influencing which genes get methylated or "turned off."[5]

The other major alternative viewpoint Sapp gives is the role of symbiosis and cooperation as evolutionary mechanisms instead of just the competition and "nature red in tooth and claw" about which we have heard so much. The decoding of DNA and the human and other species genome projects have given us a vastly different vista from which to survey the evolutionary land. We now know, partly as the result of this genome project, that species are composites, that genes intermix and jump, that all organisms are complex colonies as much as they are individuals. Symbiosis is now seen to characterize all of nature from trees to humans.[6] We know that ancient bacteria form the mitochondria in our cells and bacteria line our gut and our skin.[7] Sapp adds, "Hereditary symbiosis, taken together with evidence of horizontal gene transfer in bacteria, offers a dramatically different view of evolution from that offered by the neo-Darwinian synthesis."[8]

In this chapter I draw out some of these new ways of looking at evolutionary theory, hoping for new metaphors and new vistas on an old process. I claim that these new metaphors are not at all evidence of God hiding in the gaps, but they do make faith more rational or are even suggestive of faith if properly understood, within a wheat and tares world. I will look first at the most rapidly adapted metaphor for evolutionary theory, symbiosis or cooperation. This is not a new motif, of course, but the sense that it is a rule of evolution alongside natural selection and competition is growing.

Symbiosis

Symbiosis and cooperation works at many levels. As mentioned in Chapter 4, Sarah Coakley has worked closely with mathematician Martin Novak to reintroduce the idea that cooperation, even of a sacrificial nature, enhances survivability of a group over time. Joshua Moritz also claims that cooperation and symbiosis make a big difference to how evolution is understood. Mechanisms like symbiosis supersede natural selection to a large extent and make the theological conversation with biology more rational.[9]

Evolutionary cooperation takes many forms; in a group, stronger members might look after weaker members; mothers especially will sacrifice themselves for their young; adoption of an orphan, even of another species is not unknown in the animal world. But cooperation perhaps reaches its heights in symbiosis, where two species live together in a niche, not competing but assisting one another to survive. The evolutionary process is particularly replete with examples of this kind of cooperation. The beech forests of New Zealand, for instance, live in a delicate symbiosis with the fungi on the roots and trunks. Similarly, when parts of the Amazon were cut down to grow crops it was discovered that a similar symbiosis was occurring for these trees above the topsoil. The earth remaining after the trees were gone was almost useless for growing.[10]

Frank Ryan, author of *Virolution*, gives examples of symbiosis, both biological and behavioral. These include the case of mitochondria, which are of ancient bacterial origin. Algae and fungus are hybridized to form lichen.[11] At a behavioral level sharks astonishingly allow little fish to clean their skin and mouths of parasites and debris.[12] Many species come into various behavioral and genetic and chemical forms of symbiotic relationship, beneficial to both. Ryan says:

> [mutation] is not the exclusive source of genetic, or, for that matter, genomic variation...natural selection alone could not have given rise to the evolution of life, and its subsequent diversity, for it depends on a number of driving forces, each of which is an important source of hereditable genetic change, and without which selection would achieve nothing. Great evolutionary forces, such as symbiosis and hybridization, are of vital importance to the variation that Darwin so desperately needed, and they invoke a creativity that is very different from, and far more powerful than, meaningless static.[13]

Viruses and animals

The most extraordinary story of symbiosis, however, is that of viruses and animals. Virus fragments, it seems, have had both a tragic and a helpful role to play in evolutionary development of animals. Virus derived fragments, for instance, may constitute 9 percent or more of our human genome.[14] They appear to be responsible for the creativity and fluidity of genomes, and may well be the reason that speciation and rapid change are able to happen at all.[15]

Retroviruses are responsible for 9 percent of the human genome. A further 34 percent is derived from retroviruses in the form of so-called LINEs and SINEs.[16] These virus segments and derivatives are not the junk DNA they were once thought to be. They seem to be responsible for the ability of genomes and segments of DNA to duplicate, transpose or hybridize and fuse with other segments and genomes.[17] At other times the once exogenous (externally sourced) viruses themselves become fully functional within the genome. Viruses lose their motility and become immortal within the animal genome—a form of symbiosis. These virus segments are used actively as genes in human and animal bodies, for example to form the placenta in placental mammals, and proteins throughout the body, especially the brain and hormonal organs.[18] Every day, says Ryan,

> in the vast proliferation of virus-host interactions that are taking place throughout the biological world, wholly disparate genomes—the genetic codes of what amounts to different kingdoms of life—are fusing, virus-with-host, on every twig and branch and leaf of the tree of life, from the humblest bacteria and Archaea to clown fish, sea slugs, the eagles that soar over mountains and oceans, the monkeys, apes, gorillas and orangutans that inhabit the primal rainforests, and the primate, *Homo sapiens*, which though it has long dispersed from the African cradle, yet shares much the same evolutionary inheritance.[19]

This extraordinary window into the ongoing picture of evolution adds mystery. We might ask, yet again, why or how are the evils of plagues and epidemics used to infect the genome with viruses that may in the end aid animal and plant development. Viruses have always been seen as the least of the biological world; they may end up being responsible for the way animals have evolved. Viruses are seemingly random, yet apparently able to hitch their wagons to the great tree of life. In this sense they are intriguingly consistent with the wheat and tares found everywhere else in the history of life.

Symbiosis gives us hints of the great web of interconnectedness and fragility of which the living world is constituted. Survival is not a lone individual or even a species affair, but involves a closely integrated web of species together. Today we humans are more aware than ever of our dependence on trees, which are so much under threat in the great forests of the world, and our dependence too on the humble bee, which pollinates our food and is rapidly declining in numbers worldwide—almost certainly because we in turn have undermined its habitat. It hardly needs to be said, however, that this fragile interdependence is much closer to the image of the kingdom of God than are the previous images and pictures of competition to the death.

Life Is at Home in the Physical Universe: Patterns and Architecture of Life

Many people will have noticed that shapes and intricate patterns are everywhere—shells, the shape of trees, the ripples in sand or pond, the growth of fungi or mineral deposits, the contours of hills. All of these are a part of the exquisite beauty of both the living and nonliving world. What is less evident is the way this imprint of patterning and shape has guided life and left its imprint there in many of the exquisite forms of skin patterns, hive architecture, cell divisions and so on. Here I examine the work of Phillip Ball who—with Ian Stewart—looks at evolutionary constraints and pattern imposition at the level of physics and chemistry and mathematics. In the three books, *Shapes, Branches* and *Flow,* Ball explains many of the mysteries of life and the way in which life flows into and out of physics and chemistry.[20] This is an expanded and more subtle version of the works inspired by D'Arcy Thompson's 1917 book *Growth and Form,* which highlighted Fibonacci series and spirals, ubiquitous in nature. In *Branches,* Ball explores the visual similarities between trees, lightning bolts, and river tributaries, and many other natural branching patterns. The Pythagorean, he says,

> ... would have been less surprised than we are to find spontaneous regularity of pattern and form in the world—five-petaled flowers, faceted crystals—because he would have envisaged this orderliness to be engraved in the very fabric of creation.[21]

Honey bees build hexagonal hives and zebras have their stripes because of the physics and chemistry working beneath the surface, Ball argues, rather than because many different hives—or markings—have been tried and only one survived.[22] In this thinking we see glimpses of the way in which life "flows" within the fault lines already present in the physics of the cosmos, and expresses these patterns in a multitude of living symphonic variations. Thus the hints we discern of the interconnection of all things in the picture of the stardust that constitutes the building blocks of life and of our bodies is expanded and multiplied by these new visions of what is going on in the basic anatomy of life. These insights break some of the captivity of the old pictures and give us a glimpse of the "deeper than Darwin" landscapes beneath the surface.

Thinking especially of snowflakes Ball says, "complex pattern and form is often generated in processes that take place significantly out of the thermodynamic equilibrium..."[23] Patterns of cooling combined with surface tension produce the dendrite structures seen in living *and* nonliving branching patterns.[24] Water, essential to all life, imposes a hexagonal structure on the emerging snowflake.[25] Snowflakes are proof that life is not the only highly ordered phenomenon, but that "within the snowflake" there is "found an echo of the shapes of trees and flowers, ferns, and starfish: patterns that hint at an exquisite conciliation between geometric purity and organic exuberance."[26] Fractal patterns interestingly describe branches, mineral dendrites, bacterial growth and sprawling city boundaries alike.[27] Perhaps life, says Ball "is one of Swenson's inevitable ordered forms, waiting to burst forth as soon as the universe gets a chance."[28]

That this leads to very different views of life and its inevitability and its situation in the great scheme of things is evident. I quote Ball at length:

> [Scientists] have tended to think that, because even the most primitive organisms are hideously complicated, and because their ingredients seem to be rather rare in an abiotic (inorganic) environment, life on Earth was a remarkable stroke of luck. That, however, sits uneasily with geological evidence suggesting that life probably began on our planet the instant this became geologically feasible—that is, once the surface was not longer molten, and water had condensed from the atmosphere as oceans. Moreover, the fact that life seems to thumb its nose at the second law of thermodynamics, creating order rather than succumbing to randomness, has long left scientists uneasy...The notion of maximum entropy production and the concomitant drive toward nonequilibrium order potentially removes such paradoxes. *It implies that life itself is a result of the*

seemingly irrepressible tendency for order to crystallize away from equilibrium. If that is right, we have little reason to fear that we might be alone in the universe.[29]

In *Shapes,* Ball examines lattices and geometric order, in the leopard's pelt, the honeycomb, the bands on the angelfish and zebra and in the rippling of sand. The common physics behind all these phenomena tells us that "[the] far more complicated forms of animals and plants are created by a progressive division and subdivision of space."[30] He argues that the "how" questions in biology require a better answer than just genetics. Nature will choose, he says, "to create at least some complex forms not by laborious piece-by-piece construction but by harnessing some of the organizational and pattern-forming phenomena we see in the nonliving world."[31] Stripes and forms and patterns on animals and in cell division all follow similar deep physics and chemistry.

We can readily see that all of this amounts to a much fuller picture of evolutionary progression, one much more connected with the physics and chemistry of matter, one producing intricate patterns and revealing as it were the deeper mathematical underlay already present. Although this is a very different view of life than one focused only on the gene and natural selection, Ball is certainly not pushing any particular metaphysical barrow, and especially he is not pushing a religious one. His trilogy of books, however, do make more evident the nestling of all life, vegetable, animal and human, within the "embrace" of earth, water, air. It makes more coherent what Elizabeth Barrett Browning suggests that "*His embrace Slides down by thrills/ through all things made.*"

Similarly, Stuart Kauffmann, is well known for his work on self-organization in some molecules and the part these play in the evolution of life:

> Much of the order in organisms, I believe, is self-organized and spontaneous. Self-organization mingles with natural selection in barely understood ways to yield the magnificence of our teeming biosphere. We must, therefore, expand evolutionary theory.[32]

Emergence theories, described briefly below, are also related to these ideas. We could be held captive by these images too, thinking that all of life is *just* physics and chemistry. That is not the point that Ball is making, nor the one I wish to argue; rather I am showing that life can be rationally considered to be "expected" and that its coming to be is written deep into the fabric of the pre-existing universe. Physics and chemistry are a lot more productive, and a great deal closer to life than once we supposed. There is no suggestion that any of this makes life mechanical at heart. Rather life is not a surprise visitor or an alien in the nonliving world; indeed the boundaries between life and nonlife might well be more blurred than once we thought. The universe itself might well have more of the characteristics of life than we imagined, as Rupert Sheldrake has recently suggested.[33]

Constraints and Convergence: The Return of Teleology?

> From this perspective the impoverished world picture which the western world has been busy painting with a meagre palette of predominantly browns and greys on a scruffy piece of hardboard (rescued from the attic) might not only be re-illuminated, but in this new blaze of light the wonder might become deeper—and the risks clearer.[34]

Convergence, along with *evo devo*, gives us hints of the purposefulness of the evolutionary process, and the needless way in which all metaphors of purpose have been eliminated from evolutionary discourse. Convergence, as Simon Conway Morris describes in his book *Life's Solution* describes the evidence for invisible constraints that help guide the evolutionary process toward fitness niches, even perhaps those niches of culture and of God.

Somehow evolution "knows" where to go, knows where to look for new ideas and new levels with their attendant laws. We might not know how, but Conway Morris likens the evolutionary search to the Polynesian advancement through the Pacific; they had a way of proceeding against the tides and currents so that those same tides would return them home if they found nothing. What they were searching for, of course was land. The land was there but hard to find in the vast expanse of sea.[35]

Whether or not the evolutionary process is constrained in hidden ways has been discussed ever since Darwin's day, when even the mechanisms of evolutionary mutation were not known. For if constraints or more dynamic principles are at work these are consistent with teleology in a way that natural selection alone never has been. Convergence, therefore, is a second modality—along with symbiosis—which can be seen to change the evolutionary landscape. Conway Morris has made convergence—together with the metaphysical possibilities it entails—his life's work. He argues that evolution seems to pick up the same solutions over and over again—the camera eye in cephalopods, vertebrates and cnidaria (small marine animals), for instance. Convergence happens at every level from the molecular to the plane of intelligence; corvids (crows), dolphins and primates all fill the niche of intelligence, for instance. Other examples of convergence include the parallel emergence of leaves in mosses, liverworts, lycopods, and the parallels in ferns' roots in lycopods, horsetails, and ferns. There is a morphological similarity between the extinct Mesozoic marine reptile Ichthyosaurus and the living marine mammal Phocaena—the porpoise and Delphinus, the dolphin.[36] Behavioral convergence is found in tool use, megadonty, symbolic culture and bipedality.[37] Conway Morris sees convergence as evidence of the constrained nature of the evolutionary process. These constraints appear to be enormously subtle and deep, perhaps infinitely deep as Haught proposes of the "deeper than Darwin" influences. He is notable for insisting that teleology should not be removed from evolutionary descriptions.[38] Simon Conway Morris, for one, believes that:

> The heart of the problem... is to explain how it might be that we, a product of evolution, possess an overwhelming sense of purpose and moral identity yet arose by processes that were seemingly without meaning. If, however, we can begin to demonstrate that organic evolution contains deeper structures and potentialities, if not inevitabilities, then perhaps we can begin to move away from the dreary materialism of much current thinking with its agenda of a world now open to limitless manipulation.[39]

Conway Morris uses convergent evolution[40] to argue that as yet unknown mechanisms guide evolution toward specific goals, like sentience. He and others argue for a much greater interplay between the environment and genetic change. He continues,

> Evolution may simply be a fact, yet it is in need of continuous interpretation. The study of evolution surely retains its fascination, not because it offers a universal explanation, even though this may appeal to fundamentalists (of all persuasions), but because evolution is both riven with ambiguities and, paradoxically, is also rich in implications.[41]

What is interesting about this passage is that he could be speaking of Genesis. Both Genesis and evolution reveal their depth in their capacity for almost limitless deeper knowledge. From *Life's Solution* Conway Morris went on to give both the Boyle Lecture in 2005 and the Gifford Lectures in 2007, placing himself at the centre of the interpretive exercise in reading text and nature. "If the watchmaker is blind," says Conway Morris, "he has an unerring way of finding his way around the immense labyrinth of biological space."[42] He argues also that "evolvability itself is a selected trait."[43]

> The identification of hypermutation and site sensitivity, especially in response of pronounced environmental changes, strongly suggests that, when the cards are dealt across the table of life, aces, and kings appear with alarming frequency.[44]

The concept of convergence can even be widened enormously to matters not previously thought to be instances of convergence. Indeed in a book edited by Conway Morris a number of biologists extend the idea of intelligence to whole species, to ant colonies, and most intriguingly to plants and trees.[45]

What difference does convergence make? Conway Morris is writing as a Christian, but others involved with the convergence movement are not. From a faith perspective, however, if evolution is constrained, then the constraints work as hidden force fields, making what appears on the surface random much less so. Although the connection between convergence and *telos* (being directed toward an end) is debated it is at least plausible to make this connection. This shift in evolutionary theory makes less oblique the presence of God in evolutionary history. As a new biological emphasis, however, convergence does not instantly solve all problems of theodicy—as indeed neither does cooperation—for waste and carnage and predation remain along with these other more positive constraints and pressures. In this chapter, however, I am concerned only to show that the positive drives of evolution are numerous and complex and dynamic, and rooted in the earth from which life has come, and that in this light it becomes coherent and rational to speak again of discerning God in nature.

Evo Devo

Embryos develop rapidly from egg to organism and this development is purposeful. Biologists have asked what is guiding development and whether similar mechanisms could be hidden within the evolutionary process as a whole. This is the thinking behind the new field of *evo devo*. What guides the embryo to maturity may also give us clues to what guides evolution to a given solution. Sapp describes this uncharted area in evolution, and the disagreement that has existed between geneticists and developmentalists because the former were reluctant to concede that the cytoplasm played any role in heredity and evolution. "Ontogeny"(or the development of an organism from a fertilized egg) after all, says Sapp "is controlled and purposeful; evolution, according to neo-Darwinism, was stochastic and random."[46] In the new field of *evo devo*, the insights that come from the observation of the very purposeful development of the embryo from egg to organism, and the naturally self organizing abilities of some proteins and processes are applied to the mysteries of evolution.

Biologists have long debated how much loss and how much gain of information has occurred in the evolutionary process. The new *evo devo* theory examines closely the wisdom that developmental biology can glean for evolutionary theory, leading to the surprising result that most animals have *much common genetic material* that is simply expressed differently in the wide variety of types and species in the world. Something guides the way in which these common genes are turned on in sequence and something similarly guides the pattern of gene expression in different species. Philip Ball defines *evo devo* as being "based on the notion of a toolbox of patterning genes that as well as being activated at different times in the growth of the embryo, have been enlisted to serve different evolutionary functions."[47]

Significant differences between animals are in part just a difference in which genes are expressed at different times of development and in which manner. Jeremy Taylor, for instance, in his book *Not a Chimp*, details the way in which humans differ from other apes primarily in the degree of gene expression for brain chemicals and pathways. Sometimes the differences are a factor of ten. From one perspective this makes us more different than the difference in genes would suppose—Taylor's point.[48] From another perspective the underlying links are then emphasized—the same genes but with different degrees of expression. Gerd Müller, writing in *Nature Reviews Genetics*, says:

> Although evo devo is widely acknowledged to be revolutionizing our understanding of how the development of organisms has evolved, its substantial implications for the theoretical basis of evolution are often overlooked.[49]

Different animal designs reflect the use of the same old genes, but expressed at different times and in different places in the organism using the ancient and stable Hox genes. Sean Carroll describes an interesting picture. Ernst Mayr predicted that a search for homologous (similar) genes in any population or between populations would be quite futile. "Contrary to expectations of *any* biologist," says Carroll "most of the genes first identified as governing major aspects of fruit fly body organization were found to have exact counterparts that did the same thing in most animals, including ourselves."[50] Carroll also speaks of a basic "tool kit" of genes that "govern the formation and patterning" of bodies. He goes on to explain that the diversity we see in the evolutionary tree of life is "not so much a matter of the complement of genes in an animal's took kit, but, in the words of Eric Clapton, 'it's in the way that you use it'."[51] *Evo devo* reveals how diversity is obtained through a relatively modest numbers of genes, explaining in part the surprisingly small range of genes in widely different plants and animals.

Sapp describes how Darwin was also intrigued by embryology. He thought it provided proof of evolution. "If each species had been created independently by divine inspiration," he argued, "then one would expect that the route from egg to infant would be direct. But embryologists had found exactly the opposite... Embryos of land living vertebrates go through a stage of having gill slits."[52]

Philip Ball argues that all this enormously enriches the overall picture of evolutionary development, and that it in fact "silences some of the objections to evolution that fundamentalist religious critics have raised."[53] Evolution has tinkered with old solutions to find new ones. The evolutionary ship is so cleverly fitted out that "small changes in the way existing genes were regulated and activated gave rise to structures that were modifications of old ones and yet at the same time had entirely new functions."[54] Evo devo juxtaposes the purposeful nature of development and asks questions about why such a purposeful process would be the result of a longer evolutionary process that was of a fundamentally different character. Asking these sorts of questions turns out to be very fruitful in terms of understanding the way evolutionary change is going.

As already mentioned, a related mechanism of evolution now very widely discussed is epigenetics. There is widespread interest in the idea that we may be able to influence our children's or grandchildren's characteristics—that they might inherit in some manner acquired traits. Epigenetics is thus related to *evo devo* and to symbiosis and other newer understandings of the genome. Animals still can't give birth to children who have better physique, because their parents were physically active, but in subtle ways diet and environmental agents determine which of our genes or gene copies are turned on or off, possibly not just for the next but for subsequent generations, and not always for good.[55] Some see the rise in autism, obesity, and allergies as epigenetic phenomena.

Epigenetics, *evo devo* and convergence all give hints of the "deeper than Darwin" aspects of evolutionary theory and throw up the possibility of new metaphors for guiding our thinking in these areas. As Simon Conway Morris argues, teleology cannot be eliminated or discarded from this picture. While teleology strained against the grammar of natural selection alone, it is not at all a foreign or difficult concept in the light of convergence, *evo devo*, and epigenetics.

Emergence

Finally I raise very briefly the huge area of emergence that is now widely discussed in a variety of fields. Theories of emergence have begun to change the homogeneity of the evolutionary picture. Emergence is that process by which new more complex properties come out of a less obviously complex substratum, or in which new characteristics emerge.[56] In strong theories of emergence it is postulated that these new phenomena are irreducible to and unpredictable at lower levels. Higher levels emerge, nevertheless, by natural laws that might be hidden. An obvious example is the way in which carbon crystals take on new properties such as hardness and sparkle as diamonds that are not properties of graphite. Any scientific appraisal of diamonds must treat this level of being carbon as distinct from that of other forms of carbon. And diamonds, of course, partake in a whole cultural world that graphite does not, and vice versa. Weaker theories of emergence agree that more complex states exist but not that they are irreducible in theory. Either way, emergence encourages us to see reality as layered and to respect the top down causality of the whole upon the part. Thus the discourse possible and appropriate at one level no longer works at a higher level and vice versa.

Emergence is very evident at all levels of life; intelligence and personality emerge from the zygote even though the exact point at which these properties emerge is impossible to demarcate. Once emerged, however, intelligence has a profound downward effect upon the biology of an individual. In the evolutionary process humans can indeed be seen to have unique characteristics of wide ranging linguistically mediated intelligence and imagination and these characteristics have led in recent human history to our material and artistic culture and our highly developed moral and legal codes. The precursors of this uniqueness are evident in other mammals, corvids and dolphins but these creatures have not passed—we assume—a mysterious emergent line that makes us human. Nevertheless, all animals live within a way of being or "selving" that constitutes their particular type of intelligence and habits and these in turn have an influence upon lower levels of biological existence. In all sentient animals this intelligence itself would have a causative role on the direction of evolution. Emergence gives us a view of the universe that has multiple levels of causation.

Jacob Klapwijk has written one of the more important recent books on emergent evolution. He is more concerned with the way in which higher levels of organization participate in laws appropriate to their level than he is in arguing for irreducibility. He says, for instance, that "world history is not just the monotonous report of random events but also the overwhelming story of emerging worlds, disclosures of meaning, and realizations of new norms."[57] He reformulates evolution as "descent with innovative modification." Not merely change but innovation is characteristic of each newly emerging level or state in the evolutionary tree. This process is irreducible in the sense that it cannot be predicted or understood merely in terms of lower level laws, because the higher level participates in its own special so called idionomic laws.[58] He likens this to music. We can analyze music at the level of physics and harmony but no concert review is of this nature, nor is most appreciation of music articulated in terms of physics and harmonic intervals. Rather music is a level above physics and partakes of complex rules of composition and expression that are appropriate only to that art form, as well as some that are appropriate to others.[59] Music only exists because there are minds like ours to create it and listen to it. Klapwijk refers to each of these states of being as the *umwelt* of that particular life form.[60] The significance of this is that in a sense the laws accompany the coming to be of this new state. But in another way of seeing the laws were already there, awaiting the coming of the life form appropriate to this state. He says,

> What I have in mind are orderings that acquired, after hundreds of millions of years, a different configuration because they proved to be receptive to new, above physical ordering principles. [61]

The idea of idionomic laws is very similar to Conway Morris's use of the concept of niches. Klapwijk argues that we all know this instinctively, because we navigate life as though we can distinguish one level from another, judging and measuring by the appropriate law. These determine our behavior and our institutions.[62]

There is an interesting resonance here with a new book by Carol Kaesuk Yoon, *The Clash Between Instinct and Science,* in which among other things she mentions people who have had strokes and can still name and identify nonliving objects but not anything living. Even the brain, she argues, is hardwired for different levels of existence.[63] Importantly Klapwijk describes emergence as being anticipatory in the sense that newer levels are anticipated in lower ones, giving the impression of purpose even if that purpose is not known.[64]

Conclusion

There always have been and are now emerging variants of evolutionary theory that are more dynamic and that emphasize the multiplicity of levels at which life is working—physics and intelligence for instance—the impact of the environment back into the genome and onto behavior, and the degree to which cooperation has always been a strong element in survival and thus in evolutionary progression. None of this presupposes God. But if there were a God, especially a Trinitarian one, we might expect to find that life is compatible with and perhaps even imbued with cooperation, that there might be a more dynamic relationship between "dust" and life than a strict Darwinism might suggest, and that newer forms might emerge surprisingly from lower ones we thought we had known exhaustively. These heterodox theories leave more room for intelligence of even the individual animal to be at work in the wider process.

Biological science itself may now have reached a point where evolutionary theory is undergoing paradigmatic change. While the process of succession and common descent and adaptation is not challenged, nothing is any longer settled with regard to process, and new genomic discoveries have often been surprising. To biologists these may be seen as merely modifications and tweakings of the theory, different ways of describing what is going on, especially in a context where challenges to evolution are exaggerated in an ongoing ideological battle. What is a minor paradigmatic change to biology, may, however, have profound theological consequences. Indeed, as described previously, a few biologists have also begun to challenge the nonteleological framework of biology.

None of this evidence is compelling in terms of God being there behind it all. God is still hidden but not, I would argue, completely obscured. I do not think that belief can endure for long if we are saying that we believe in things that are completely obscured, that we must believe in spite of all the evidence to the contrary. Elizabeth Johnson, as mentioned earlier, has said of history, the "divine presence and action in the world are not so intangible as to leave no discernible historical traces."[65] Something similar can be said of God's presence in the natural world. At the very least, to believe in the present climate we have to be clear that we are responding to God in nature and in Scripture, that together there is a case to be made. After all when Job questions God it is this evidence from nature—from animals in particular—that is given to him. Without nature it is hard to trust our other intuitions, it is hard to believe that we aren't kidding ourselves, or living in bad faith. This is why I think that we must point to a constrained, mathematically driven evolutionary process, which is nevertheless contingent and random in part, and conclude from the perspective of faith that this is a radically new way of understanding the evolutionary process. This understanding of evolution—unlike a hard neo-Darwinism—shows hints of the deeper subtle processes at work beneath the surface, and makes sense of the emergence of genuinely new levels of being. This deep subtle interlocking beauty and order do speak of God's presence if we have the eyes to see and the ears to hear. The grammar of these processes is more consistent with the Scriptures, which speak of being knit together in our mothers' wombs (Ps 139:13). Why does this matter? It matters because it breaks the captivity of the picture that was in so much tension with that of Scripture, providence, and experience, and gives a multitude of different vistas. Klapwijk, like Midgley, talks about our having been "hypnotized...by the dream of a one-dimensional evolutionary continuum in nature."[66]

If the power of a loving God is evident in creation we can more easily trust God that the less desirable aspects—the tares—will be eliminated, healed, or brought to fulfillment. We don't have to know exactly which aspects are less than desirable. We can be agnostic about the source of all manner of gradations of evil. It also means that evolution is not being propelled forward so to speak *only* by means of a competition to the death, which is the depiction that is so much at odds with the law of love. That means we are not in evolutionary competition with the animals, for instance; rather mutual dependence and interconnectedness are typical of the whole process.

The mystery of life itself is suggestive of the constant upholding of being in a dimension at "right angles" to our present existence. Thus God works from below and more actively to uphold all that exists, without in any sense cancelling the freedom, nor the apparent randomness and indeed historical contingency of parts of the process of evolution. What matters to theology here is that the process is not at heart one of randomness, driven only by the ever pressing will to survive and dominate others. There are deeper, more loving, gentler aspects at work that can be seen to set the direction of the evolutionary process. Life and animal passions and personality are at "home" in this universe. This makes more possible a reconciliation of evolutionary theory with theological impulses that affirm the salvation of humans and of creatures as the *telos* or end of history, and also a theology that emphasizes the peaceable kingdom, the end of war between and among the animals.

John Haught, as was mentioned in the previous chapter, has urged us to see that evolution can be seen in this multi-layered way, that scientific explanations need not exhaust the possibilities for explanation of a biological event. Nevertheless, offering explanations at several different levels only works if the explanations do not in some sense contradict one another or cancel one another out. Natural selection by blind mutation has always had a grammar very different from that of the indwelling *logos* and life breathing Spirit. Newer understandings of evolution may break this deep incompatibility and open up a way to seeing formal and even final causes as a not incoherent way of understanding the evolutionary process. In this way, although theology is not directly adding to science, it is contributing to the sense or meaning of the picture science presents.

If we were to imagine a divine being undergirding and enfolding all that is, we might expect this picture, however it all began, of one layer after another of coherence and cleverness and fit. Evolution is a process constrained and funded by the physics and chemistry of the world, giving life to nature's patterns, resulting in small cells that have grown with purpose into nature's life forms, and able at times to respond quickly to environmental pressure. The process enables a permanence that allows long-standing species but incorporates also an accompanying creativity that leads to speciation. All of this is the remarkable phenomenon of evolution. Some of these dynamics break the captivity of a picture of random mutation and selection alone that has long dominated our collective minds.

9

Dualism or Tares in Evolutionary History?

Affliction makes God appear to be absent for a time, more absent than a dead man, more absent than light in the utter darkness of a cell. A kind of horror submerges the whole soul. During this absence there is nothing to love.

SIMONE WEIL

CRITICS WILL POINT to the seeming dualism that a wheat and tares model presumes. In such a model do not good and evil battle each other throughout history as ancient Zoroastrians and Manicheans and Gnostics have always thought? I would argue that the particularly Christian way of looking at our human predicament is not dualistic but that Hart is probably right when he speaks of a "'provisional' Christian dualism in the New Testament."[1] This provisional dualism is consistent with a wheat and tares world and with forms of causality and influence that are more subtle than those with which we are familiar.

The Christian story always has been told that what was intended for evil or to harm life has been used by God for good (Gen 50:20). This is the suggestion of a world in which both crucifixion and resurrection happen, in which the end has already been told in the beginning. This grammar of final redemption is very different from that of a simple dualism and is in some way also the best description possible of sovereignty. Sovereignty is not that God is more powerful than anything else in the universe, which must in some simple sense be true, but that God has a more subversive power over evil, and moreover, that humans are to be recruited into this subversive power. The other aspect that differentiates Christian wheat and tares from another dualism is that we are to fight the evil, though not with the usual weapons of war but by means of the weapons of the spirit

(Eph 6:14). Thus healing, prayer, practices of worship, love of enemies—and I would add, love and protection of animals—are all the grammars of the gospel, the force of which overcomes evil and will overcome evil. For some reason or other we must go through this struggle, but for Christians there is a way ahead, which is the path of resisting evil.

How then does this wheat and tares understanding apply to the animal world and evolutionary processes and the suffering of animals? First of all it means that God must not be seen as affirming and endorsing *all* that happens in the evolutionary process. Nor must God be seen as using these evils as means toward an end (as in the best of all possible worlds), even if God does promise to bring good out of evil in the long run, which is subtly different. Thus the wheat and the tares gives us a way of understanding the evolutionary process that separates good from the evil and allows us to praise God for what is good. The tares, however, are as mysterious as they are ubiquitous. Does a wheat and tares world accord at all with the rest of Christian theology, most of which is done with only human life in mind? I look now at a number of biblical and theological scholars whose work is very readily seen to be compatible with a wheat and tares human prehistory even though they are not directly speaking to this.

N.T. Wright, Resurrection, and Evil

The conclusion of the wheat and tares parable has the final harvest of the wheat in which the long sought separation occurs. The resurrection has always been the evidence and the support for the final victory of the wheat over the tares in history. In Chapter 3 I mentioned resurrection as a way of responding theologically to evil, especially if it can be extended to the whole of life. There are progressive Christians who say that resurrection is what we experience now, and it is true that the Marxist critique of religion necessitated our insistence that the *basileia*, is both here *and* to come. Christians are not *only* looking for life after death; rather the grace of God in Christ, as Kathryn Tanner describes, enables us to be truly human here and now.[2] The other extreme, however, which denies the historical resurrection also, to my mind, denies the ubiquity and the depth of evil, an evil that may extend deep into evolutionary history. Tom Wright, for instance, sees the resurrection as evidence that the solution needs to be radical.[3]

Tom Wright does not write with a scientific or evolutionary point of view in mind, but the conclusions of his long scholarly indwelling of the text of Scripture in dialogue with the nuances of the *zeitgeist* is wonderfully open to this perspective. In *Evil and the Justice of God* Wright affirms that evil must be taken seriously, much, much more seriously than we are inclined to take it. He also acknowledges, in a manner in keeping with the wheat and the tares that "the line between good and evil runs through us all."[4] We get evil dreadfully wrong when we ignore that fact and try to divide the world into those who are tainted with evil and those who are not.

Wright also deplores the separation of philosophical views of evil from theological ones. He affirms that "[t]he Christian belief, growing out of its Jewish roots, is that the God who made the world remains passionately and compassionately involved with it,"[5] a position that is of course consistent with Habel's emphasis highlighted at the beginning of Chapter 6 where he affirms that the "story of creation embraces the presence of God as an integral part of the cosmos, perpetually moving within the cosmos."[6] Yet this passionate and interacting God is not acting in a way we understand. Nevertheless, the Scriptures attest to God's presence, and to the plaintive voices of generations before us who have believed and yet been puzzled. Wright thinks that ultimately there is no answer to evil that we understand, except that God has come to rescue us and also through and with us to rescue the cosmos. In spite of the warning against trying to understand, Wright writes a book, as I am doing. There is a subtle difference between trying to map out in too much detail the territory and limits of evil and acknowledging its presence and the importance of thereby not downplaying its effects.

True to Wright the solution in so far as there is one, is to place the problem within the wider story of Israel's redemption. He affirms that which we can see and experience if we are honest, and names this the "providence" of God.[7] He gives as examples the providence Joseph son of Jacob brings to the story of Israel, where yet for many decades no sign of God was visible or sensible.[8] He affirms that "God will *contain* evil," and God "will *restrain* it."[9] Even before we get to the New Testament Wright sees in the Suffering Servant passages, the signs that God will redeem all life by sharing in it.[10] In the book of Daniel, Wright argues, we have an image of the Son of Man making war against the beast; this is, he writes, a little like "Adam in the garden being set in authority over the animals."[11] It is about creation being "restored, put back into proper working order."[12]

Wright draws more traditional conclusions when he says that "the evil that humans do is integrated with the enslavement of creation."[13] Yet what of the creation before humanity, we ask? In the latter half of the book Wright sets out his reasons for *Christus Victor* atonement. He urges also a renewed sense that there are indeed principalities and powers and suprahuman powers that influence and control and permeate reality as we know it. Wright senses that the problem of evil is a much, much bigger one than has hitherto being admitted. For those of us interacting with animal life and evolutionary history we can assent to this but also give even more depth to that conviction. Evil does needs to be overcome. The evil that touches the whole of creation and is intertwined within it is crying out for a solution. Wright also points to the political powers and deeper darker forces "which operate at a suprapersonal level, forces for which the language of the demonic...is still the least inadequate."[14] I would add that if these forces are at work then it makes sense to imagine them at work in some manner in the whole of creation. I will return to this theme in the following chapter. Here it is sufficient to show that this modified dualism still makes most sense of our existence and of the existence perhaps of all matter and certainly all life.[15] Evident also is the fact that although the good and evil are intermixed and the line runs through each one of us, nevertheless, we know unequivocally that God is against evil. God's ultimate judgment against evil is being withheld precisely because the good must eventually be unravelled from the evil. Yet in the death and resurrection of Jesus there is evidence that God's salvation will be sufficient. "The taint of evil lay heavy on him throughout, and somehow he bears it, took it all the way, exhausts its power."[16]

The wheat and the tares, and operations of evil in relationship to good are two sides of the same coin. One brings us inevitably to the other. The other inescapable reality is that the connection between the good and the evil in our lives and those of animals remains hidden and secret, open to multiple interpretations. Yet the goodness is never completely devoured by the evil, however brutal it is. Public notices throughout Germany proclaim that "We will Never Forget." We might ask why is there a temptation to forget? We forget because all history falls into the hands of the interpreters, and interpretation is always open to truth and to deception. The deeper and more complex the reality the more open and more hidden is the meaning.

Jon Levenson and Evil

In *Creation and the Persistence of Evil*, Jon Levenson argues that we should not explain away the presence of evil.[17] He says of H. Richard Niebuhr:

> by removing that boundary (between light and darkness) Niebuhr sets up a new one, the boundary between creation and moral

struggle. No longer is God set against the darkness, even if only by reputation. Instead, he affirms everything because he creates everything.[18]

This is exactly the problem we discover in many solutions to the problem of evolutionary evil. The boundary between light and darkness has been removed. And I would include here even Christopher Southgate's solution of intratrinitarian kenosis to the extent that it affirms all things as the best of all possible worlds. Levenson says of biblical monotheism, that it

> ...refuses to attribute value to everything that exists. Some things exist that ought not to, and these deserve to be blasted from the world. Not everything that exists in nature is good or conforms to God's highest intentions. Some of what is, is not yet good.[19]

Levenson's vision is consistent with a wheat and tares world, though in this case the evil should not be blasted from the world, until the end, until the final harvest, even if we might pray that it is sooner. Levenson says:

> The central affirmation of the apocalyptic is that the evil that occurs in history is symptomatic of a larger suprahistorical disequilibrium that requires, indeed invites, a suprahistorical correction. As evil did not originate with history, neither will it disappear altogether *in* history, but rather *beyond* it, at the inauguration of the coming world.[20]

Levenson looks at Ps 33:4–7 and connects this with Exod 15 and the creation:

> the essential unity of creation and exodus, two great acts of deliverance wrought by the gracious God on behalf of his powerless allies. This is another instance in which cosmogony and redemption or, if you wish, myth and history, or creation and revelation, work in tandem and not independently of one another. The very endurance of the world offers testimony to the boundless benevolence of its author, his indefectible resolve that the dark forces of chaos not triumph.[21]

In a similar vein Levenson goes on to talk about the "control" of evil in Gen1:1–2:3, a control brought about by the ordering power of a speaking God. Like God we must bring order and distinctions to a world of chaos and evil:

> Among the many messages of Gen 1–2 is this: it is through the cult that we are enabled to cope with evil, for it is the cult that builds and maintains order, transforms chaos into creation, ennobles humanity, and realizes the kingship of the God who has ordained the cult and commanded that it be guarded and practiced.[22]

Thus Levenson, too, provides a biblical interpretation that is consistent with a wheat and tares world of nature. Levenson, like Wright, also understands that humans are to be with God the co-redeemers of creation, overcoming evil and chaos. Thus animals and humans are not distinguished by one being the domain of the wheat and tares and the other not so being. They are distinguished by humans being in relationship to animals in a redeeming and ordering world.

Goldingay and the Snake

Goldingay also has intriguing things to say about evil. He argues that Scripture uses a number of different images to describe evil, the "snake, the dragon, the accuser and the Devil." Like other theologians he is caution of giving any one image ascendancy, or relating one to the other too quickly.[23] Concerning the snake in Genesis 3, however, he says this:

> The snake has the brashness of Tiamat, but also a subtlety she lacks. In the process of creation there was resistance to God taking the form of a frontal attack, but it failed. There is now resistance taking the form of discrete innuendo.... Innuendo succeeds.[24]

Goldingay goes on to describe the snake as standing for "dangerous, deceptive, dynamic power outside of God's sovereignty, or seeking to do so."[25] These are extraordinary and quite far reaching statements, though the consequences of what is being said are not followed up at all. Could there have been a frontal attack on God in the process of creation? Where did this come from? Who else was around? Was there really a power seeking to involve human beings in its overarching power? Is there indeed some modified Christian dualism? The implications of Goldingay's statements are certainly consistent with some challenge to God's sovereignty, which is deeply embedded in the fabric of creation and could easily be experienced by us as the tares amongst the wheat.

Hiddenness and God Transforming Nature

Nevertheless, in the domain of parables there is always depth and mystery, and hiddenness. The parable of the wheat and the tares ends with the injunction that those who have ears will hear or listen. They will see. Others will not. A wheat and tares world is one that can easily be misinterpreted. Whatever the disturbance in nature, whether it be shadow sophia, or evil by another name it is ultimately beyond our understanding. The unease, the mystery, the pain remains. The ends are left untied. Robert Jenson has explained some of this mystery by appealing to Rom 8:20 and to the passage about the creation being subject to futility. He sees the lack of *telos* (end or goal) or ending to be the frustration to which nature is at the present moment subjected—giving hints of promise but not yet delivering, poised as it were between ending and pointlessness. "Just at this point," he says,

> . . . we may think of the devil *and his fallen angels*. For if the angels are in the inner-creaturely reality of creation's direction to God, then demons must be precisely gaps in created teleology, junctures where creation's dynamism can skid out of control.[26]

Scripture is more interested in directing us to God's power and to God's ultimate appropriation of all alien power. Although wheat and tares give us a justification for recognizing signs of God in a world touched by evil, nevertheless God is more often than not still hidden. If God is present in the child Jesus and yet did not prevent the slaughter of the innocents, and if God is present in creation and yet creation is troubled and divided then God is hidden indeed. Much attention has also been given to this hiddenness. Philosophers have argued that what Bob Russell (after John Hick) calls "epistemic distance"[27] is needed to give make genuine knowing and believing possible. The truth cannot be forced upon us, either of truth of God nor of the creation; hiddenness is a necessary part of reality. Swinburne argues that this distance is required to provide the possibility of moral goodness.[28]

The problem of evil returns. It might be the case that evil is "explained" by these suprapersonal powers, but why, if God is so close, so entwined with the world does this presence not do more, especially post-Easter, to stop the progression of evil? We might see glimpses of the "glory" but why so little else? Scripture has never spoken of a God who is there to be seen like any other object or person; nevertheless, "ever since the creation of the world his eternal power and divine nature, invisible though they are, have been understood and seen through the things he has made"(Rom 1:20). We should expect to see God, to see God's imprint in the natural world, but never to see fully or wholly. Although Scripture speaks of a God who has acted, and is discernible at some level by signs, these signs are deeply embedded in the natural, in much the same way as the encounter at Emmaus was embedded in an ordinary meal of bread and wine.

In the science and theology dialogue a Trinitarian God allows us to affirm the ontological density of divine presence, whilst nevertheless also affirming God's otherness and transcendence. The Trinitarian God coheres with a wheat and tares universe, one in which God is present but also apart, and in which the freedom of other agents and even of evil is not obliterated or penultimately subjected to the purposes of God. Trinity and enspiritedness, even the presence in creation of the Christ, makes possible the supposition that matter is tinged with divinity, and that we should expect that the dense levels of structure and emerging layers of organization glimpsed in science know no end. Evolutionary history gives us a paradigm of the emergence of the surprising and the radically new; each level of consciousness and complexity giving way to hitherto unprecedented novelty. A Trinitarian immanence is also consistent with our knowing of the universe, which tends to be both detailed, and at the same time on the edge always of something seemingly beyond us.

Theology should lead us to think that even the vastness of what we have so far glimpsed is only a thin slice of unfinished reality, rather than a broad outline overview of all that there is. This sense of possible unlimited surprise is affirmed in the continuing output of scientific research. That surprise and novelty and newness should typify our experience of reality is consistent with Iain McGilchrist's argument that the right brain senses this novelty and freshness while the left brain, the dominant hemisphere in our civilization, sees reality as mechanical and ordinary.[29] Novelty and newness also typify the Peircian category of firstness.[30]

The temptation of science, however, is to think that we are within sight of the end. Palaeontologist, Simon Conway Morris, speaks against this hubris when he argues that there is "no limit to the complexities of the world we inhabit." He hopes that this understanding might "refresh our wonder at Creation."[31] Barth, too, suggests something similar when he says, "it is to be noted that the revolutionary discoveries of recent decades show that nature even as at present constituted may hide unsuspected mysteries and possibilities of further development."[32] An ever-opening world is therefore a hint of God. The world is not yet full of the glory of God. The world is not yet new. God does not yet dwell among mortals (Rev 21). Do we suffer and do animals suffer because the Trinitarian God is with us and not yet fully with us? Is the kenotic explanation finally the reason for suffering? I would argue that it is not God's absence that is the explanation for evil; rather it is that God's presence is not yet fully overpowering evil in the way that Rev 21 predicts. The presence of both wheat and tares obscures the glory of God and requires a moral discernment of what is going on. This discernment has been radically undermined in a hyper-Darwinian world. The absence of God, however, is not in order that we be free, for a case can certainly be made that humans are most free when God is most present. As Robert Jenson and Karen Kilby and others have suggested the grammar of freedom and God does not work like that of humans with other humans.[33]

The Supernatural within the Natural

The theological response to Darwinism from the beginning has always been: where is there room for ongoing Divine action by which our religious intuitions affirm God? Believers want there to be a place where they can look at the created world and say "that action bears the trace of God," as Moltmann has suggested in pleading for a new natural theology.[34] One response lies in Schleiermacher's analogy with Christology, and his argument that there is finally no clear distinction between the natural and the supernatural, the rational and the suprarational, that God is encountered in the man Jesus Christ, and in analogous other natural ways.[35] In the end Jesus Christ, as the Son of God incarnate "extends an influence and a redeeming activity" that is unlike that of any other human, and this is a new work of God in him, but he does it within the confines of human nature, which means, Schleiermacher argues, that "there must reside in human nature the possibility of taking up the divine into itself, just as did happen in Christ."[36] There is a crossing of the boundary; there is divine action, but at any point the consequence is noted, it is observed within the natural continuum. The work of God in Jesus was done in a human who was one of us and was not observably or "scientifically" different from other humans. Jesus obeyed the Father in perfection but was also humanly free. Jesus would have been a man with retroviruses and with the stamp of an animal past in his genes. Jesus was not merely programmed to act out the Son of God role but acted freely in dynamic relationship with the Father and in the Spirit. The supernatural was very much embedded in and indiscernible from the natural.

A similar case may be made for recognizing the Spirit within the creation. Nothing is ever as we expect it to be; there is always new and hidden depth of structure and connection. Intuitively we discern patterns and structures that confirm the presence of a beauty that precedes us and surpasses human construction, and a beauty that is rational and alive and even mathematical in form, like the glory at which David Hart hints. The world resists ultimate explanation and reduction even though exponentially increasing amounts are known. A sense of purpose pervades the universe, even though explicit definition of this purpose evades us. Some of this purpose is evident in the increasing levels of emergent evolution, described so well by Jacob Klapwijk and Philip Clayton. These levels require discernment, however. One might draw an analogy with the three dimensional random prints that, if stared at long enough, reveal stereoscopic images and a vitality not seen if only the flat pattern is perceived. Nature, too, can be perceived that way, as flat and klugelike or as emergent; the emergence of new levels of genuine novelty betray purpose, while the flatland of naturalism is consistent with Dawkins' "blind pitiless indifference."[37]

Thus the picture of God and of nature that is implicit in the biblical and theological tradition is neither that of a God who can be read off of nature, nor of a God who is completely obscured behind it. God is revealed: "the heavens declare the glory of God," but God is also obscured, the signs are never unveiled. Furthermore, nature itself is the locus of God's action. Christ became human, a part of the stuff and material of the universe, and did signs and miracles within and out of his human nature. Scripture pays little attention to the miracle of the supernatural becoming natural, and much to the incarnational revelation of God in Christ. We might see analogies in nature in the way in which the parts contribute to the whole without causation, and the way in which gravity is "caused" by the geometric parameters of the universe. Terence Nichols has referred to something similar as contextual causality.[38]

Conclusion

In this chapter I have used the work of several biblical scholars who affirm the depth of evil that is portrayed quite bluntly in Scripture and the way in which a wheat and tares world coheres with the picture they present. Lastly I have examined the hiddenness of God and our ongoing perplexity in light of God's only partial presence.

We are thus enabled to look at the very mixed picture we have of animals, their predation, their dying, their suffering, and their extinction and not to be surprised, but neither are we compelled to affirm the mixed picture as good. While some animal suffering will be explained, as it is in humans, as the best of all possible worlds or as a necessary teaching in order to avoid harm, or even as character building, much of what happens with increasing frequency with higher orders of animal sentience is something that prefigures human sentient suffering. Much of what humans experience is humanly derived suffering; human history, however, is only a blink in evolutionary history. If we know that the world is inexplicably both good and bad, perfect and corrupted, and that telling these apart is often fraught then we can understand animal suffering in the same way that we might understand human suffering, as a part of the corrupted world, caused in some sense by the evil known variously in Scripture as the Evil One, powers and principalities, or "shadow sophia." This evil is eventually to be overcome. The story of redemption is one that covers the whole universe in its groaning.

The affirmation of a wheat and tares universe brings immediately to mind the question of fallenness. I have hinted thus far at a progressive fallenness, and this, along with dominion, I examine in the next chapter. I argue that at the dawn of humanity a new potency is given to both through the powers of the human being. This can be seen as fallenness, but a fallenness that is necessary if humans are to come to full sentience in a world in which a principle of evil is already at work, especially in the animal world. Such a fallenness explains why evil seems to originate with humans—because our minds are so extensive and our powers so great—but also explains why fallenness was not the beginning of evil and why animal suffering is on a continuum with human suffering.

10

The Fall and Beyond

For I am convinced that neither death, nor life, nor angels, nor rulers, nor things present, nor things to come, nor powers, nor height, nor depth, nor anything else in all creation, will be able to separate us from the love of God in Christ Jesus our Lord.
Romans 8:38–39

IN PREVIOUS CHAPTERS I have argued for a wheat and tares interpretation of evolutionary history. In viewing our long history in this way I am able to say that not everything we see in the evolution of life and in the lives of animals is necessarily good nor intended as God's perfection; the goodness is mixed with the bad. Thus there is no need to say that God intended the cruelties of predation and extinction and the perishing of the weak because these *may* not have been what a loving God intends at all. In arguing for this interpretation I am also assuming that there has been a continuum of evil throughout the history of life, even if new levels of evil have been achieved through the efforts of human animals. I have argued that states of affairs don't have to be intentional to be wrong; they can simply "miss the mark" of what God intends. This may not get us very far, you might say; for by this means we enter the territory of cosmic enemies of God. I argue that affirming both the wheat and tares allows us to get beyond the idea that God has caused or intended cruelty, or parasitism or extinction in a direct way, and to get beyond the idea that things just are, or that they are the best of all possible worlds, or that they are a necessary evil to produce higher creatures. A wheat and tares grammar avoids having to pronounce everything that has ever existed as good just because it is.

The question still remains: if we have overthrown the idea of a complete rupture of life in the fall of humanity does that mean that we should overturn also any idea of human fall? Is the discourse around fallenness outmoded? I argue in this chapter that there is still a way in which it makes sense to speak of a human fall and that human fallenness disturbs not only our relationship with one another but our proper relationships to animals, and that it may also make sense to speak of a cosmic fallenness. If loving dominion is one of the main functions of our humanness then one aspect of fallenness upends this, making us vicious and uncaring of one another and also of our animal charges. There is an inversion of our intended role. If this is so, I argue, then the redemption achieved in Christ must include as part of the gospel, a return to a care of animals, to bringing out the best in them. Some aspects of dominion will be considered in this chapter and the consequences for animal ethics in the concluding chapter.

The Meaning of the Fall

The Fall was once imagined—at least in the West—to be the sin of Adam and Eve, who accumulated in original sin a debt on our behalf at the dawn of humanity.[1] Treated historically, the fall was said to bring humanity and the whole created world from paradise to tragedy. In a sense this doctrine has also accentuated humanity's difference from animals, for animals are imagined to be the passive recipients of human mistakes. Moreover, when human behavior is bad this is observed as "fallen" or "animal" behavior, thereby projecting onto animals, human faults. Within the Christian tradition, humanity has also taken upon itself the huge burden of being responsible indirectly for all evil, because in the past—in the tight story around death and fall–evil could be ultimately explained by the "fall." Evolutionary theory changes this. In the words of Van Huyssteen, "it is no longer possible to claim some past paradise in which humans possessed moral perfection, a state from which our species somehow has 'fallen' into perpetual decline."[2] At the threshold of humanity something did emerge, but that something, that spark of divinity and aggression, preceded us, as did a natural world, itself fallen in some respect.[3] How then can we imagine the fall? The fall of humanity is indeed, as twentieth century exegetes have insisted, part existential, part mythical. Rikk Watts and other Old Testament exegetes have shown us the ways in which Genesis can be liberated from some of our more rigid salvific stories.[4] With John Goldingay and Henri Blocher I would suggest that the story of human fallenness must have an historical/etiological edge.[5] Goldingay, as mentioned in

Chapter 1, reimagines the fall as the refusal of the first humans to take up a new calling to take the world to its destiny. Blocher argues that while Genesis has a highly complex literary and mythical structure that does not thereby rule out *some* form of historical reality and *some* historical reading as well:

> Such a combination of imagery (of whatever provenance) and a message about definite events is familiar in Scripture; one need only think of Ezekiel's allegories, of apocalyptic visions, and of many of Jesus's parables. It involves no tension. It should cause no embarrassment. Thinking otherwise is unwarranted prejudice…The problem is not historiography as a genre narrowly…but correspondence with discrete realities in our ordinary space and sequential time.[6]

If we allow an historical/realist *edge* to Genesis this immediately places us in conversation and potential conflict with science.[7] Even when this narrative is no longer intended, ambiguous liturgical and theological references to Adam or Eve often throw up the suggestions of this history in some form or other. Without this historical edge we run into the same problems we do when we insist *imago Dei* is to be understood only in relational or functional terms, and not in a substantive paradigm; the past is then imagined in scientific terms but not in religious ones. Is it possible to think that the very expansion of human intelligence magnified the violence and predation that was already present in the prehuman hominids out of which *homo sapiens* emerged? One theological interpretation of all this is to see the fall as an inevitable result of an expansive and curious and aggressive hominid endowed with enough inner drive to make the transition to language and speech and dominion of the whole earth.

The fall would then be a progressive matter, one that occurs and reoccurs through history, because the traits that make us human also make us vulnerable, and as the perhaps fallen nature of animal predation takes on a life of its own in humanity. The fall is the engagement of new human powers of mind with idolatry and temptation and hubris. In this fallenness, human animals come under bondage and experience brokenness and evil in an unprecedented way, requiring but also making possible the salvation that Christ brings. This progressive fall, however, does not entail quite the ontological separation from animals it might once have done. It does not suppose a state of perfect paradise preceding human becoming, and it does not suppose a causal relationship between disease, aggression and this fall, except in the sense that humans became powerful instigators of both good and evil in the world. This perspective allows for more continuity in the development of *homo sapiens* from hominid precursors, and more realization that a great deal of what we once saw as fallen is a part of the natural order, which may in itself be fallen. Then it is also possible to ask whether there are degrees of human fallenness. Michael Polanyi has asked whether, in the turn to technology, we have not taken again from the tree of the knowledge of good and evil.[8]

If we return to Genesis 3 and to the story of fall it is interesting, as Goldingay and many have noted, that there is a serpent already there, at the heart of the story.[9] Some part of the created order does in fact always represent temptation. A part of our history has humans making an alliance with a preexisting evil, or with some alien power, or with some aspect of creation that is out of kilter. There is an element of temptation and dissonance and hubris at the heart of finite created reality. A component of temptation and the possibility of deception accompany humanity's new almost godlike powers. It is interesting, too, as noted in a previous chapter, that temptation is central to the story of Jesus, and frames his life from the wilderness to Gethsemane. The serpent, then, does not cancel out the goodness of creation, but the serpent is there, speaking words of "wisdom," and is able to lure in the human in a way that nonspeaking animals cannot be lured.

Whatever the truth about the fall, as human forebears made the transition to humanity, we know that they did not come to the threshold completely innocent. Nature itself, although perfect in some way, is also itself already fallen or at least incomplete before humanity arrives. Andrew Linzey is a strong advocate of the fallenness of the natural world as it is. He says: "The concept of the fall...constitutes a composite rejection of the idea that the creation as it is now—is—at least in this respect—God's original will."[10] To deny this truth, says Linzey, is to say that parasitism and predation are neutral or positive aspects of the natural world, or worse that they are to be emulated. This means that we can affirm the coherence to theological concepts like *imago Dei* and fall, while, nevertheless, insisting that the separations that are endemic in our way of thinking are not fully justified, biologically or theologically. Humans are the same but different from animals. We are very much the same as one another. Humans share with animals the tragedies of the world indwelt with temptation, and live in a similar but different culturally and linguistically enhanced bondage.

As humans we all share the same violent and creative and inquisitive past. We have come out of the animals and we need to attend to that fact. Where once we might have interpreted all imperfections or incompleteness, all tragedy, all evil, as stemming from a fall from paradise, that is no longer possible. Humans, however, have enough to answer for without taking on all the burdens of evil in the world, burdens that Christians believe God has already borne. Understanding different levels of fallenness allows us to imagine salvation for the whole created world perhaps linked in some way to the salvation of humanity.

In the mid-1850s there was a debate that anticipated ours in many ways. On the one side were those who held the world to be quite static—having been made perfect 6000 years ago, and having fallen into depravity, and full of humans who were awaiting redemption; on the other side were those who espoused a more evolutionary theory that posited long aeons of time before humanity, and who therefore saw the universe in a much more dynamic and progressive way. True there had been no perfection in the past, but nor were we stuck with what we have now—humans and God could work together for something better.[11] The dominant late twentieth century hyper-Darwinist position, however, is neither of these. We have come under the sway of a much more pessimistic evolutionary philosophy, entailing a much more absent god if any at all. The evolutionary process is seen as a mixture of randomness and law in natural selection. This process has no personal intentions, no concern for humanity above other creatures, and is associated with no final or formal causes. The hardening of the biological sciences under the influence of the Central Dogma and neo-Darwinism brought about this change. I discussed the biological aspects in Chapter 8, but we need to heed these nineteenth century differences because evolution in itself can give rise to very different views of the world and cosmos. The nineteenth-century progressive dynamism may no longer be accessible to us, but the view of a world in which the wheat grows—as well as the tares—is at least possible.

C. S. Lewis and the Problem of Pain

In some ways the clearest of voices is still that of C. S. Lewis. Lewis had a particular affection for animals and engagement with the problem of animal suffering. In the past, animal suffering, and indeed all suffering and imperfection, could be blamed on the human fall from grace. Rather than dismissing all notions of fallenness, or reinterpreting fall in a purely existential sense, Lewis argues for a theology of fallenness that is wider than humanity in scope and origin; humans cannot be responsible for all the disorder in the universe because animals existed for so long before humans. These animals preyed on one another and suffered pain—they became extinct, he argued, long before others made these arguments. "The intrinsic evil of the animal world," says Lewis, "lies in the fact that animals, or some animals, live by destroying each other."[12] Taking note of animals as sentient beings, capable of suffering and not just experiencing pain, leads immediately to questions about God's goodness and about animal salvation. These questions are resolved if one postulates that animal predation is just God's way of doing things. Humans must live by a different code, they say. Lewis disagrees with this assessment, as do I.

If animal predation and suffering is a problem, what is the theological solution? Human suffering is resolved at one level by noting the benefits of suffering, and the grace of being one with God through Christ. Humans are compensated by a relationship with God and through eternal life. Human freedom is also a theodicy defense. But animals have no similar freedom of choice; they do not know good and evil as we do. God could not have intended this. Therefore something outside of God caused this state of affairs. This cause, according to Lewis, is the fallenness of angelic powers. He says: "the doctrine of Satan's existence and fall is not among the things we know to be untrue: it contradicts not the facts discovered by scientists but the mere, vague 'climate of opinion' that we happen to be living in."[13] Lewis goes on to cast disdain on the "climate of opinion" as a guide in any moral matters. He does, however, want to maintain the distinction between humans and the creaturely world, and this he does by affirming that animals can be resurrected only if they are the tame animals of humans. Lewis argues that it was a part of the original mandate for humans that they restore the lack or fallenness in creation, but he believes humans failed to take up their mandate and instead joined the enemy. Lewis introduces the problem of animal suffering and of the evolutionary repercussions of the fall for a possible human cause, and then he presses toward animal redemption. In his fairy stories animals are very much a part of the redeemed Narnia, but he leaves the details and the hard lines of the doctrine open. Lewis then has laid the ground for a contemporary understanding of fallenness, though we might ask, knowing more of our genetic and probably behavioral inheritance as we approached the species boundary, whether a "human fall" away from God was not inevitable, given what came beforehand. Lewis was also working within a much more restricted zone of knowledge of animals. He was a friend of domestic animals, and these therefore held a special place in his theology of salvation.

Torrance and Linzey

Two other theologians interact with and extend Lewis's argument. The first is Thomas Torrance. Like Lewis he believes in a prehuman angelic fall, and specifically because he takes the scientific depiction of "nature red in tooth and claw" seriously. Torrance understands the fallenness in the natural world as disorder between its different levels, and the human calling, he argues, is, together with creatures, intended to bring out the hidden surprises of nature, something science can surely be seen to do. Where he extends or makes more explicit the arguments of Lewis is in the subtle interplay of different levels of existence, and in defining the work that humans and the creation were intended to accomplish. Explaining some of the mix of redemption and tragedy that is the human predicament he says:

> However, when man himself is seized of evil, and his interaction with the Creator is damaged and disordered, his interaction with nature becomes damaged and disordered as well. Something very different takes place, for the whole balance of nature is upset. Man continually infects nature with his own disorder even in the midst of these priestly and redemptive operations.[14]

For Torrance there is an added level of fallenness that comes to pass when humans are involved. From this stems much of the suffering of animals at human hands. It does not explain, of course, animal predation or extinction many aeons before the emergence of mammals, let alone humanity. Torrance goes on to say that it is in humanity rather than in nature that evil "has lodged itself."[15] And it is in technology that the redemptive and destructive powers of humanity have been magnified. The powers of technology he applies also—as does Jacques Ellul—to the subtle regions of abstract thought.[16] Drawing on this deep sense that humans can bring enormous good and fruitfulness to and through the created world Torrance articulates more formally than does Lewis, the notion that humans may have been intended to bring peace and harmony back to the levels of creation; instead humans have perverted and enormously magnified the evil.¹⁷ There is something very attractive in Torrance's view, and yet I would want to be very hesitant about a theology that almost presents a human fall as a surprise for God. Nevertheless, Torrance does bring out the sense in which humans exposed the corruption harboring in the creation, enormously magnified it, and also made possible the coming of an incarnate messiah.

The second theologian has devoted his life work to animal theology and ethics. Andrew Linzey adapts and interacts with the insights Lewis applied less systematically. He agrees with Lewis about the animal pain, and about the prehuman fallenness of nature. (Drawing on the work of Michael Lloyd, he does give other possibilities—the fall of the world soul or the gradual fall of creation as in process theology—but he dismisses these as inadequate.)¹⁸ Linzey affirms Lewis's use of imagination to do theological exploration—in the image of the Great Lady,¹⁹ for example, who keeps dogs, cats, birds, and other animals, tending them all and enabling their kind interconnection. In this context one might note the particular perfidy involved in raising an animal to inflict violence, as happens all too often these days. It is not a harmful act done against the neutrality of owning an animal but a perversion of our intended relationship with animals, which is that of loving dominion, bringing out the best in them. He describes how humans become more human and animals more their God-given selves when they interact with one another.

Where Linzey expands Lewis's arguments is in saying that if cruelty to animals is wrong, then so is eating them. Humans can now make a difference in reversing demonic corruption by themselves electing to kill and injure as few animals as possible, or at least, as argued in Chapter 11, by taking great care with living animals and killing them humanely. We are thus able to see Lewis's contention that animals can only be understood in their relationship to human beings as a deeper issue for practical theology. Humans are morally at the center of creation: as their fall affects the nonhuman world, so too will their redemption. Since animals are involuntarily tied to human sin, the redemption of humanity matters to the animal world. Linzey applauds Lewis's theological orthodoxy and the way in which he keeps alive the drama of redemption and the crucial place that animal/human relations have in that drama, while nevertheless steering clear of the deification of nature or of pantheism.

Linzey also expands Lewis's arguments about self and resurrection. Lewis argued that it was only possibly in the development of a self in relationship to human masters that an animal could hope to achieve salvation. Linzey, however, argues that recent scholarship in the area of animals makes it evident that there are higher levels of sentience very close to that of the human already existing in animals.[20] Many others have ventured into this field. No one of course, knows exactly what a redeemed life for animals would be like, or whether particular animals or just representations of them are present, whether they live in the mind of God, or in a more real way, if there is a more real way. But then of course nobody knows what the redeemed life means for humans either.

New Fall Arguments

There have been many other recent attempts to reconcile Genesis with an evolutionary understanding of the fall among conservative theologians for whom the biblical text is more in focus than evolutionary history. They are motivated either by a sense that the fall is a doctrine too central to be easily given up or—as I would also argue—by the sense of tragedy that pervades Genesis and appears too central to the entire sweep of biblical narrative. Randy Isaac makes a case for progressive fallenness.[21] Thus he sees a fall begun by humanity but increasing through the years with increased rebellion. Although I think this understanding of fall is insufficient, not taking into account prehuman violence, it does make the concept of the fall more fluid, and opens it up to the possibility of happening in stages or over a long continuum. Nik Ansell attempts to draw a connection between the call of wisdom in Proverbs 8 and the "voice of the serpent" in Genesis 3.[22] Satan, in Ansell's understanding is not flesh and blood, and is not a fallen angel, but rather is an aspect of creation that "becomes increasingly distorted by the growth of human sin until it becomes a power that is totally opposed to the coming of God's Kingdom," becoming an "active reality that is external to human beings."[23] This is an attractive idea that does satisfy some of the evolutionary problems and also the intuition we have that sin leads to further sin. Where it is weakest, though, is in dealing with the long history of prehuman brokenness, the long prehistory of the human race. And is a hypostasis of creation that breaks away really any more satisfactory than a fallen angel?

A recent collective look at fallenness is addressed in a book by evangelical scholars, *Darwin, Creation, and the Fall*, published to coincide with the Darwin anniversary in 2009. The burden of most of these articles—as mentioned briefly in Chapter 1—is that Scripture must come first, that science is tentative (though in every case there is an admission that evolution has happened), and that there is no true conflict between science and faith if looked at from the right perspective. Sam Berry and T. A. Noble end by saying that "Darwin did not cause the sea of faith to ebb, because faith does not depend on knowing how the creator has worked."[24] I would argue, with Berkhof, Moltmann, and others, that we do need to see something, and we do need some level of coherence because there are alternative beliefs, and what we do see is very puzzling and bewildering.[25] Darrel Faulk, writing in this vein, chides Darwin for not realizing that his agony over evil was not new.[26] Faulk, however, does not bring the science into focus; he does not dwell on the extent to which Darwin discerned that he did provide an alternative to origins that was a very powerful justification for atheism, and that his theory brought to light the dark side of a process that appeared to be of Divine provenance. Sam Berry, similarly, considers the issues of Romans 5 and death but does not mention the link between natural selection, the survival of the dominant individual, and depravity. Instead he considers the possibility that the initial act of defiance "spread" in the same way that the bestowed imago did.[27] T. A. Noble sits very lightly with even the extent of physical death in the evolutionary progression, though he acknowledges the tension.[28] Death and Romans 5 is a common theme in these articles, and several authors are willing to concede that a mythical Adam can powerfully and rhetorically be juxtaposed to an historical figure without loss of integrity in the biblical witness.[29] These authors also argue that death means primarily spiritual death, citing Eph 2: 1–5, Col 2:13, 1 John 3:14, John 3:1–14, Luke 15:32 as examples where spiritual death is obviously intended.[30] Equating death in Romans 5 with spiritual death alleviates hermeneutical scruples that many who have taken Scripture seriously have with the evolutionary process.

In the articles by and about Blocher there is real interest. Blocher does want to maintain an historical fall, claiming with Ricoeur that if there was not an historical fall then God is responsible for evil. I would argue that this otherwise very nuanced article does not give full credence to the extent of violence in our cousins, the chimps, and does not dwell firmly enough on the links that must exist between violence and the way we were made in our evolutionary history—in a process indwelt by God. Thus it is interesting that he ponders when the first humans might have emerged—perhaps the Cro-Magnon at the dawn of agriculture—but for him animal extinction and even predation do not loom large on his radar of suffering. When animals can truly be shown to suffer Blocher does worry, but not enough to budge his theology.[31] Richard Mortimer carries on this discussion in dialogue with Blocher. Mortimer very usefully brings out the role of the covenant in his and in Blocher's thinking, especially in light of Romans 5.[32] Whatever the fall was, it involved some covenant, some agreement with God, under which this new species came into a different relationship to God, one in which judgment, sin, grace, and salvation all suddenly made sense; there are resonances here with Klapwijk's insistence that new levels of evolution participate in new and particular and relevant idionomic laws.[33] While the idea of covenant might seem to get us somewhere in terms of the meaning of human fallenness, it should not be forgotten, as John Olley points out, that the Flood story has God making a covenant with *all* flesh.[34] Blocher and Mortimer, however, make an interesting and compelling case for including covenant in our understanding of the fall, but the flip side of this covenant is a relationship of trust and care for the rest of creation—a role for humans as co-redemptors with God of the animals.

In summary we might ask why bother with a doctrine that is now outmoded and that has been proven to be false in its more literal sense? In the end it comes down to a theological discernment that humans are caught up in evil that transcends human wickedness and that some aspects of the complex web of life and its development are completely at odds with the command to love. Fallenness—first of the animals and then in a special sense of humans—makes sense and diminishes both those predicaments, so why discard it, and why discard the element of the demonic when the Scriptures are so full of it? A theology of fallenness also enables us to say "no" to the idea that suffering is necessary or a part of God's kingdom in some way.

Christopher Southgate on Fallenness

Not everyone agrees, however. Christopher Southgate argues strenuously against a doctrine of an historic fall. He rejects the idea of a prehuman fall, seeing this move as arbitrary and arguing that those who accept this position do so for reasons of anxiety to maintain the *status quo* in theology. I would make the counterargument that affirming fallenness, especially prehuman fallenness, takes us way beyond the domain of "safe" religion from a mainstream perspective. I am also very distrustful of most Christian affirmations of the demonic, as I mentioned in Chapter 9. The ontology of evil cannot be so neatly dealt with. Nevertheless the conclusion of the Scriptures is certainly that something else is going on.

Jesus spends a great deal of energy in his earthly ministry battling Satan in the wilderness or rebuking him in the words of Peter, healing people of the demonic, and praying that they be protected from the Evil One (John 17:15). Is this all just cultural difference, a primitive metaphysics, or a way of speaking about something real? Or were *all* these afflicted and healed people just mentally ill? Moreover Jesus's miracles are not just those of healing but also affect the storm and the number of loaves of bread available for consumption. Thus his healing power is extended to the very core of the way things are. Paul indeed promises us, as Southgate affirms, that the whole of this groaning will one day be healed. An affirmation of prehuman fallenness, if by that we mean that something is awry that precedes humanity, is not arbitrary, but biblical; it is not easily culturally relativized. Indeed the inexplicable nature of the evil to be overcome also makes sense of the contemporary rediscovery of *Christus Victor* in the atonement.

Nor, I contend, is affirmation of fallenness motivated by anxiety. One does not get into the midst of this particular quagmire of science and theology by being overly anxious about orthodox doctrine of a particular kind. To affirm prehuman suffering and tragedy of any sort is already to have taken steps that will, in the end, affect deeply every area of systematic theology—our personhood, human freedom, the nature of God, our view of the end, and of the beginning, and of the work of Christ. The time for anxiety is long past, and it is not the motivation; I would argue in the defense of fallenness in C. S. Lewis, Michael Lloyd, Andrew Linzey, or Thomas Torrance.[35] Moreover, fallenness need not signal a fall within the mundane limits of the era of earthly life, but instead may point to an origin outside the limits of our story. The great advantage of fallenness is that it is biblical *and* it helps to explain the depth and unevenness of evil and malice within the human domain, but perhaps also within evolutionary history as well.

Murray on Fallenness

Michael Murray surveys the motivations for a fall defense of God's goodness, and the fall as an explanation for evil. He explains that in the biblical revelation there is a pervasive sense that "something about our world seems more deeply and radically askew—something that can only be explained by a corruption that has eroded the integrity and thwarted the flourishing of the natural order on a cosmic scale."[36] He points to the lucidity of the fall arguments throughout the Middle Ages, where it was argued that death is privation and therefore evil and therefore linked to Adam's Fall. That the fall, which now appears so grotesque in many quarters, made eminent sense is evidenced by the seamless unity of the story that has Adam and Eve at the beginning, the great images of restoration in the prophets, and the accomplishment of redemption in Christ in a cosmos filled with angels and demons and spirits of all kinds. This all came to a crashing end with the Enlightenment and then with Darwin, because, as Murray points out, not only did Darwin show the extent of pre-Adamic pain and suffering, this suffering was the very means of creation itself.[37] The fall separated God from evil; Darwin linked God closely to it all or eliminated a Creator altogether.

The most interesting objection to a fall defense that Murray presents is one hinted at by Bishop and Perszyk and mentioned in Chapter 3. This is the so-called fragility problem. Why does God create in such a way that the fall of humans—or indeed of prehuman beings—creates such problems for the rest of creation? Surely God could make a more robust, more independent, less fragile world?[38] Murray frames it this way: "...God created things so that the integrity of the natural order was, in some important sense, initially *dependent upon* the integrity of the moral order."[39] How does the fragility work? How does a fall affect the rest of creation? This objection really does get to the heart of fall discourse. It offends our very independent individualistic consciousness that one bad deed could so easily affect another. Yet as we delve deeper into the physics of entanglement and the interconnections of the ecological balance, and indeed into human fragility in matters of attachment and abuse, it does seem plausible that these connections are ubiquitous. There is a fragile coinherence of all in all. If all is working well there is nothing more beautiful; if there is disruption and pain the interconnection is devastating. The book of Lamentations can be read as an exquisite and tragic example of this interconnected web of fragility and oppression and suffering.[40] Thus I believe that although we might imagine a more independent, robust world we can also see the beauty and necessity of the one we have and of the need for fragility as the other side of love.

Related to the fragility problem is that of incoherence. Murray argues, especially if the fall is a fall from paradise, why would two sentient creatures in paradise turn away from the good in the first place? Others cast doubt on whether creaturely freedom is worth all the pain and suffering that a fall has brought about. Murray considers along the way the counter-factual defense that there is no pre-Adamic pain and suffering. He also examines whether what we now call evil was perhaps not evil until humans came along.[41] Thorns and thistles, for instance did not matter until humans were trying to farm. There is a limited truth to this, but the focus of this book—and his—is the admitted reality of animal pain and suffering. Murray then turns his attention to similar defenses, which assume an old universe and prehuman pain and suffering. He notes that Isa 14:12–15 and Ezek 28:12–19 have both been used to bolster the idea of a prehuman fall of Satan.[42] He describes how on this defense:

> ...sin, pain, and death were present before the advent of human beings, having been introduced as a result of an angelic Fall. Human beings later came on the scene with a redemptive task: to "rule over" and "subdue" a created order which had already been set on edge.[43]

This, says Murray, "has all the explanatory advantages of the traditional fall CD [*causa Dei*] for natural evil without the ahistorical liabilities."[44] Murray himself is not convinced, though he does claim that a fall defense is coherent and valid, even if he doubts that anyone would make a full-blown theodicy out of it. The objections include objections to the fall of Adam and Eve but also that of *ad hocery*. I would argue, however, that something similar is neither *ad hoc* nor unbelievable in light of the extent of evil we have witnessed—the Holocaust, the selling of children into prostitution, ethnic cleansing, the extinction of species, the change of mood between Palm Sunday and Good Friday. We might wish we lived in a world without these evils, but we do not. In such a world the existence of prehuman evil is not only not *ad hoc* but also makes sense, if indeed sense can be made of evil.

Instead Murray asks of Satan and his cohorts: "Could these beings be to blame for the fact that human beings often have bad backs, myopia, liability to cancer and heart disease?"[45] Put this way we are inclined to say, of course not. We know why these things happen. In a world that just had bad backs and myopia we might be inclined to settle for a "soul making" defense of evil, or something similar. In a world of horrendous evil and millions of years of predation and carnage, as well as human atrocity, some dark subtle tares at work among the wheat makes sense.

Fallenness for Us and Other Animals

This last section is very speculative. I imagine here what fallenness might mean to us and to other animals. I have postulated that almost certainly we did not invent cruelty or our most fashionable of current sins—sexual predation—nor did we invent mimetic desire after we crossed the species boundary. I assume this because there are other creatures like chimps and bonobos who are not fully sapient and who are our closest cousins; the former are violent, patriarchal, and status seeking. The latter are promiscuously sexual—not that that matters to the bonobo! My point is that all of these characteristics, which seem to beset human life, are possible in presapient life. Either *homo sapiens* had similar characteristics before we crossed the species boundary or our cousins have "fallen" but are not yet sapient. I assume this also because meat eating is inherently more viciously predatory than vegetarianism. The skills that allowed us to hunt are also the skills that might allow us to kill, though other factors like mimetic desire are also needed as motivation for murder. Chimps, however, have plenty of envy and could conceivably make the transition with much baggage in terms of the prerequisites for brutality.

Simon Conway Morris, Michael Ruse, and others speak of the niches or spaces animals have inhabited—water, land, air, and also the cultural/tool-using space.[46] Ruse—and Conway Morris—insist that Darwinism does not invent the niches; rather animals push or are pushed into these spaces.[47] From the eyes of faith *homo sapiens*—and perhaps other species in the past—has inhabited the God-space, the space of being made in the image of God. There is no rule, though, that says we are the only possible inhabitants of this space, either here or on some distant planet. Other animals are on its boundary, in terms of having limited culture and tool making and some level of symbolic communication. Were we to imagine either chimps or bonobos moving into the God space we inhabit they would not do so as innocents. This might also be true for corvids (crows) and dolphins as far as we know. Even if a species did come to the boundary of sapience more innocent than *homo sapiens* it would experience both the advantages and dangers of this space—a knowing of God, the responsibility of care for lower animals, and the temptation to lord it over them and other members of one's species. It too might experience another fall. It is inconceivable that any earth species could move into this space without already being tainted and showing signs of "fallenness" or without falling immediately upon entry. This species would, after all, have to contend with *homo sapiens* and would either perish or fight. Thus the burden of this chapter has been that "fallenness" can be meaningfully applied to all of life on this planet, at least to sentient life. Because this fallenness is mixed with perfection it is hard to separate out the wheat from the tares, though the chimps being on the edge of warfare, as de Waal suggests, can most easily be seen to be fallen. Even a predator playing with its catch, or a lion devouring a flourishing creature, I would contend is fallen in some way, however necessary to its survival this behavior might now be. The sudden disappearance of whole species might well be a fallen aspect of this reality, or we might one day see it within a larger context as not evil in the larger scheme of things.

Dominion

Assuming a fallenness to the whole of nature, and a special level of fallenness to humanity means we should not attempt to emulate nature in its entirety; rather we should take the responsibilities of our God-space seriously in care of the animals, in bringing out their better sides, whether they be domesticated or wild. Dominion within an evolutionary perspective must have partly connotations of ability—the power to know and investigate and image all created reality including that of animals, and almost godlike powers. Dominion also connotes the other side of these powers and their attendant responsibility—to tend and care for the garden, to recognize the covenant God has with all flesh, to meditate on our common fate and our common destiny. This along with our much greater current understanding of the place of animals in God's provenance, and our very close evolutionary ties—and ecological fragility—should give us cause to be open to new relationships with animals and to take seriously the guardianship of their niches as well as our own. Interconnectedness of all things means that this care of other niches will also entail care of humanity's own future.

A fallen but also good world can be seen to reflect God's glory when we look upon its dense levels of perfection. A fallen but perfect world removes the cognitive dissonance that comes from trying to always live above or against the means of our becoming. Such a way of living inevitably separates us and divides us from animals. The knowledge that we share a fallen world, on the other hand, gives us a sense of solidarity with all life.

11

Concluding Ethical Reflections

> *...there is no question of keeping the chimps out of the castle. They and many other animals have always been inside, and only our conceit and prejudice have stopped us from seeing them. They are all over the ground floor, which is still a central area of our life as well as theirs. But there are many other floors to which they do not go and cannot, because they have never wanted to enough, and so have never developed their powers beyond a certain rudimentary point.*
>
> MARY MIDGLEY

THIS HAS BEEN a book about the problem of evil, given animal suffering. As such, it has not been focused on ethics and on our responsibilities to animals, except to mention dominion in the previous chapter. I have argued that the long pre-human history of animals must challenge our theology in significant ways. In particular an old theodicy with hints of a past paradise must be overturned; the kernel of theology that was the traditional reading of the Garden of Eden can be no more. No more can we think that humans have driven some line between paradise and tragedy, that humans are a special creation and animals not, that the image of God emerged out of that boundary and divided us once and for all from animals, or that humans have souls and animals have none.

I have argued quite otherwise, that there is no sharp boundary between humans and animals, even though I would argue—with Van Huyssteen and Jacob Klapwijk and others—that humans are unique among animals and participate in religious and ethical and political and cultural levels that other animals do not; nevertheless, there have been other sapient species; there are continuities and discontinuities between ourselves and higher primates. We now know that Neanderthal and Cro

Magnon interbred. There was mixing and intermixing for tens of thousands of years. We did not arrive in one magical moment from what we were before.

I have argued in previous chapters that humans are animals and that as animals we share intelligence, tool making, culture, axial emotions, altruism, and also aggression, selfishness, and in some case mimetic desire with the other animals. Humans and other animals are all the results of the same very long evolutionary process and the same sets of homeobox and other gene combinations, and have similar patterns of viral inserts in our DNA. Why then have animals been so badly treated, excluding a few domestic types? Why has animal behavior been the standard for the lowest human habits; and why have humans previously seen animals as devoid of axial emotions? The answers are complex, but lie partly in our theology and partly in Darwinian pictures of a random and directionless process that has stunted our moral imaginations of the animal world.

Theology, for instance, has long drawn distinctions between the image bearer and the animal, between the human who is a part of the salvation story and the animal who is not, between the animal capable of genuine love and altruism—and also sin—and the animal who works only by instinct. These distinctions have all been associated with the paradise to fall to redemption story within systematic theology, which I have argued needs to be greatly modified. Instead if we look closely at Genesis, an extraordinarily deep and mythical document, our rejection of its literal meaning notwithstanding, we find that animals and humans share creation on the sixth day, animals and humans alike were saved from drowning in an ark. Animals and humans alike perished in the flood. Post deluge all life was brought into covenantal relationship with God. The whole creation is groaning as it awaits redemption. The peaceable kingdom gives images of restoration and peace between predator and prey. Animals appear in surprising places throughout Scripture—human and beast are the faces in the vision of Ezekiel; the woman in the Song is called by animal names, the beasts welcome Jesus and accompany him to Jerusalem.

The other story, however, that has fueled our sense of human superiority has arisen out of the Darwinian model of random evolution, together with the move toward efficiency and control that the technological world has perfected. Evolution by natural selection has become the ultimate efficient process, the means by which all the life has appeared in all its charming and beguiling glory. Only human life has risen above the vicissitudes of this relentless and pointless process, becoming wise to the underlying machine. The Darwinian picture of evolution alone, I have argued, tells very little of the truth. It does not see the pattern of darkness and light, for evolution is characterized as much by love and cooperation and symbiosis as it is the opposite; it is as purposeful as it is seemingly random. If instead of an old story of paradise and fall, or a selfish Darwinian process we see instead a tightly bound covenant of human and animal, and a process in which love and cooperation are most important then animals must be seen in a very different light; they become possible bearers of love and healing and delight, as revealing in some sense the face of God, but at other times as being in need of healing and release from bondage as are humans.

Science and theology are both in the midst of paradigmatic change. They require of us a transformation and conversion in our relationship to the rest of the world. Where once I might have thought as a child that animals had no place in God's kingdom, we now see how parochial that was. Where once science might have given the impression that humans had evolved to a point of "otherness" that is now being questioned. New dynamics in evolutionary theory and new metaphors give us fresh reason to question our connection with animals, especially those with whom we live in close proximity. Symbiosis, convergence, and cooperation as evolutionary mechanisms should help us to see these are acceptable and desirable values in the wider ecological sphere. Emergence might give us reason to believe that our pictures of the world are incomplete, tentative and penultimate. Surprises will be evident in all spheres of life and relationship.

If we place these values together with the Christian doctrine of dominion new possibilities open up; mutuality and healing might be expected in human animal connection. Domestication, which to the pure eco-activist eyes might be viewed with suspicion as unnatural and imperialistic, can be seen in the light of the deep cooperative mechanisms of evolution to be desirable and a part of God's intended order of things as well. We are seeing hints of this new outlook in the many experiments that are opening up where animals are entering into places where humans are young or particularly vulnerable—prisons and schools. Animals have long been used as companion dogs for the blind and deaf and on farms; a new sensitivity to the healing and mutual benefit of animal human relationships is beginning to be evident. I include a number of notable stories where animals and humans have seen this mutual benefit.

Dominion and Healing

One of the repercussions of a theology of dominion is that humans in their uniqueness do have a responsibility for the animal kingdom. Humans should be "facing" the animal world and being formed and informed by these relationships. At their best these relationships will be healing for animal and human, and in being healing can be the first fruits of the kingdom of God.

What evidence is there of the healing power of animals? I present here a few stories of animals and humans in deference to a new way of knowing which must emerge out of our more rationalistic ways of knowing in theology and in science.

Bekoff and His Dog Jethro

There is a story told about Marc Bekoff and his dog, Jethro. This story hints at the possibility of human kindness creating a space or overflow of kindness into the animal world and vice versa. The following is an abstract of this story:

> *The dog and human had a long and close relationship. As a younger dog Jethro came into the house holding an injured bunny in his jaws, dropped the bunny at Marc's feet and then was transfixed and ever-watchful through the long process of healing. Jethro became the bunny's closest companion and minder until it was strong enough to leap away.*[1]

Jethro must have been a kind dog, but this kindness can only have grown through the special attention and mutual love with a human for whom animals are special. That this kindness then spilled over onto the rest of the created world, subverting what might be the usual predator-prey relationship is interesting. Christians believe that grace can reform human consciousness, healing sin and inbuilt tendencies toward selfishness and aggression. This story shows that humans can also be mediators of kindness, constructing spaces in which natural animal tendencies toward predation are also cured and redirected. In this case kindness was magnified as it was shared across species, giving a little taste of the peaceable kingdom.

Dog Henry and His Autistic Boy Dale

This abstract is from the book, *A Friend Like Henry*. It is the remarkable tale of an autistic boy, Dale and his dog, Henry.

> *After his already harried parents were persuaded to get Henry—the dog, the child and dog formed an immediate bond. One day Dale's father, facing an intractable, isolated and immovable child, had a sudden insight that he should speak "through" Henry. He would pretend to be the voice of the dog. It worked. Dale's parents used this device from then on. The child was tamed and healed by the dog.*[2]

Here the parents speak of the healing the dog effected, and the older largely healed Dale also affirms that the dog was a crucial turning point in his life. The bond with the dog was immediate. The intervention of the parents, however, was the means by which this intimacy between dog and child became widened to gradually include the affections of family and the wider world.[3] Of course, there are many stories of horses healing, and horse whisperers in turn taming the wildness of horses. These experiences and anecdotes don't fit into any of the nice categories we have about the world and its proper ordering. The story of the child and the dog is a tale of mutual naming. The child named the dog and the child was in turn called out by the dog and the parents together. Naming became humanizing. The dog was the mediator, an unconditional and loving presence to the child who hovered on the brink of exclusion from the human world of speech and affection.

The Wild Parrots of Telegraph Hill

The Wild Parrots of Telegraph Hill is a widely screened, award-winning documentary by Judy Irving. The documentary tells the story of Mark Bittner, who communes at a very deep emotional level with the birds outside his small apartment. This experience draws him into a close ongoing relationship of naming and care with the flocks outside his home. Here is a synopsis of his story:

> *A down and out musician decided to take seriously the idea of accessing nature where he was. His local nature turned out to be once domesticated parrots who had freed themselves and were living in flocks in urban San Francisco. The documentary shows his growing interactions with*

and knowledge of the personality and habits and pecking order of his parrots, and how it changed him. Near the end of the film we hear the story about how he began this journey of bonding with the parrots. He picked up a sick bird one night as he read before turning off the light. He felt a wave of gratitude wash through him. As he put out the light and put the bird down he was startled by a wave of grief and desolation. In the morning the bird was dead.

All of these stories speak to a deeper connection than is possible within many of the religious and secular pictures of modernity. They don't, of course, mean that animals are never vicious or oblivious to humans, but as C. S. Lewis has said, "even now more animals than you might expect are ready to adore man if they are given a reasonable opportunity; for man was made to be the priest and even, in some sense, the Christ, of animals—the mediator through whom they apprehend so much of the Divine splendour as their irrational nature allows."[4] At the very least the pictures within which we have been held captive did not urge us to interact with animals, nor to expect that they might bring with them gifts of being and otherness and companionship. Many stories of human interaction with animals speak almost of shared love, stumbled upon in most cases. Mark Bittner of Telegraph Hill was surprised and at first ignored the emotions he experienced from the contact with sick birds; Dale's parents discovered Henry as a mediator. They tapped into Henry's very real love of Dale and Dale of Henry. Marc Bekoff did not expect his young dog to be so attentive and caring of another, much smaller, species.

These stories of inter-species transformation and healing do make sense within a worldview informed by a God of love and by an evolutionary process in which, even in a wheat and tares universe, the primary motif is love and symbiosis and care. These stories show that just as human minds are reformable by grace, so other species also can enter into the redemptive power of grace and transformation. There are many others, of course; a cat in a cancer ward has shown uncanny knowledge of who is most ill and will visit and comfort them; a friend of mine once "talked" to a pelican on a remote beach, telling it gently how beautiful it was. It came up to her, preened, leaned its head toward her and deposited a fish at her feet, and paddled away again, as though acknowledging her appreciative presence. Dolphins have surrounded struggling swimmers. Often we have ignored animal attention because *our* attention has been elsewhere.

We might compare this with the vision of the universe given by Jonathan Edwards:

> That consent, agreement, or union of being to being...may be called the highest, and first, or primary beauty that is to be found among things that exist...Yet there is another, inferior, secondary beauty, which is some image of this, and which is not peculiar to spiritual beings, but is found even in inanimate things: which consists in a mutual consent and agreement of different things in form, manner, quantity and visible end or design.[5]

In the contemporary world Edwards's vision has been lost. If we can regain a theodicy of the wheat and tares it might be possible to regain a perspective on the evolutionary process as one epitomized by love and care and by multiple embedded layers of interaction and emergence. If so we could expect that there were multiple untapped resources for humans and animals in redeemed relationships, and in the new values that new evolutionary metaphors make possible.

The Repercussions of Seeing Love at the Heart of the Created Process

If love is primary and is manifest in creation, then creation and animals, in particular, are intended to be in relationship with humans and vice versa. Healing and health and the love of God are all there to be experienced. What might follow is a mandate to increase the normal everyday interaction of humans and animals, especially for people who are in situations of sickness or vulnerability. Prisons, for instance, could become healing farms if animal-human interaction were a primary means of humanizing, healing and rehabilitating broken humans, and where animals benefit also from the much needed companionship. This is beginning to happen in prisons where inmates train dogs for the disabled with mutual benefit to dog and people. The scope and possibility is enormous.[6] Children would benefit from animals being central to schools; attention-deficit children, autistic spectrum children and those who have been abused stand to benefit most from an ongoing relationship with animals, but domestic animals would also benefit; these animals love nothing more than to be desperately needed. Schools might have animal parks and an interaction with both tame and wilder animals could become an important part of socialization and education, healing and reform. Animals might migrate from the realm of children's literature to the very centre of their lives. If farms were no longer so full of animals to be sent for slaughter more land might become available for animals of other kinds, or animals we might "use" in more gentle ways—goats, laying birds, cows and sheep. In the church we might find that animals become an important part of our liturgical life and that the story we tell our children would include the animals and their importance to God.[7] It would not begin and end with the children's Noah's Ark as it does so often now.

For too long the natural world has been an opaque or ambivalent window to God. Human relationships with animals have reflected our overall ambivalence about the creation we embedded within. If, however, animals are reflecting God's love in some way we can delight in them. Noticing them and interacting with them will speak to us of God as much in their wildness as in captivity. Animals can be seen as natural healers where we have previously only seen them in utilitarian ways or as wild and threatening. I argue that we must interact with animals to find the full breadth of God's revelation. Moreover, within the Christian church we might expect this interaction will speak to us of the love of a triune perichoretic (mutually interpenetrating and relating) God and may well be a source of healing for both human and animal.

The Wheat and the Tares

If we truly do live in a wheat and tares world, what repercussions does this have for our interactions with animals? I think it means that we are never ultimately going to bring about the restoration or the kingdom of God on earth; although we can participate in healing and in new ways of interacting with and being with animals the good will always be linked in some ways with evil. Being vegetarian or arguing and standing for the rights of animals are all a part of the pointers to the kingdom, signs of its presence and its future. Theologically we must conclude that our fate is linked with theirs in some way, as the ancient covenants suggest (Gen 9:11). Our being with and defending animals entails that humans spend some time finding out what the world of a particular animal means, what they need and how, if they are domesticated or farmed, we should respectfully interact with them or care for them.[8] In every way, however, humans alter the *umwelt* (life world) of the animal for better and for worse. We must be always trying to understand the ways in which our interventions—high rise housing, pollution, noise, elimination of habitats, and so on, are potentially harmful for animals as well as humans.

In acknowledging, however, that we live in a wheat and tares world we can see parallels with other intractable problems. Our bodies have been formed in an evolutionary process in which meat eating became dominant—hence the idea of the *paleo* diet. For some people groups like the Inuit this reliance on animal products almost exclusively was highly successful. We carry in our bodies the mark of predation. We also carry aggressive tendencies and enormous confusion over sex and sexual roles that have all undoubtedly been a part of our genetic inheritance. All of these darknesses need to be resisted in some way or other, but only a utopian totalitarianism will think that we can simply by our own wills eliminate these tendencies in human society, because so much of what we do is a part of a larger whole (we "send" armies, for instance, overseas), and a great deal is also unconscious, the elimination of one error introducing another—as for instance communism and capitalism have both done. These larger structures within which we live are what we know as structural evil, or what John Wesley called "complicated wickedness." The world has also long become familiar with the idea of unconscious sin. In the Anglican liturgy we pray to be forgiven for the sins that are clear to us and those that are not. In the end, although many victims can speak if they have not been destroyed, animals remain silent partners.

Should We Be Vegetarian?

What do these close links with animals mean for our diet? In this book I have come close to seeing predation as an evil present in the evolutionary process from the beginning, although the particular evil surrounding predation I think occurs in the fear which is induced by the stalking and slow dismembering of an animal that knows it is cornered. Nevertheless, how can I justify eating meat? I struggle with the answer to this question, and I would readily agree that a vegetarian or even a vegan diet is an ideal to which I think we should be moving. In a wheat and tares world, full of hard practical decisions and innumerable conventions and responsibilities and a meat eating culture, the ideal is not always easily managed.

Vegetarianism is similar to other ideal values we humans espouse, values like equality and liberty and pacifism. They are often not readily compatible; some countries have idealized liberty, others equality but the ideal of both is not humanly possible. Pacifism has been modified by the idea of world community policing, in deference to the on-the-ground need for minimal levels of coercion to protect innocent people. Most of the difficulty in espousing absolute values is rooted in biological realities and deep evolutionary psychology—the tendency of humans to aggression, the deep felt need for justice, the desire of all for peace.

Similarly for the ideal of vegetarianism; for some people groups our very bodies let us down. I agree that the most satisfying answer is to err on the side of caution and be a vegetarian all other things being equal. It goes without saying that any meat in our diet should be sourced from places that treat animals well and have relatively stress free slaughtering methods. While this may not appear to be possible there are a number of factors that might mitigate the horror of animal killing. Small animals killed by their owners who also have cared for them normally choose methods that are swift and as un-menacing as possible. Caring farmers who are killing their own animals will normally choose methods that are as swift and un-menacing as possible. Barbara Kingsolver has argued something similar in *Animal Vegetable Miracle*.[9] She and her family gave up eating meat when the drove past the cattle holding sheds within which many and perhaps most farm animals spend their lives in the US. When she could raise turkeys herself, however, she went back to eating limited amounts of meat and animal produce. For her the important part of the exercise was to live locally and to know that the animals, because local, were either her own, or being treated well.

That killing methods can vary is evidenced by the work of Temple Grandin. Grandin in many ways relates better to cows than she does to people, and she has designed slaughter houses where the animals are gently squeezed and calmed before killing—she had found relief for her own anxiety in a similar squeezing machine; this at least is a far cry from the terrorized animals and brutal killing methods of many freezing works. In New Zealand, where I live, most cattle and sheep have been treated well and spend their days outside on green grass. Nevertheless, there is a growing worldwide humanitarian concern for animals; there are also of course the tares growing in the form of more ruthless housing and culling of battery animals under pressure from economic concerns and the interests of mass efficiency.[10] Nevertheless we also find evidence of an increased awareness of the need to be vigilant on matters concerning farming and animal husbandry. In some parts of the world you would become *de facto* vegetarian if you were at all concerned about these matters, as Kingsolver has suggested.

Should vegetarianism be an ethical position? I certainly welcome the conversation that is now taking place—and the increasing numbers of vegetarian Christians. I agree with Linzey that vegetarianism is an ideal, all other things being equal, and that we should not ever eat meat without pondering its ethics and without being aware of its source. Even if we don't insist on vegetarianism we should all be prepared to make this an option in any public communal meals.

If vegetarianism is the ideal can we really recommend it as the normal everyday path? In Christian theology there is an aspect to the healing of the planet that will come from outside, by the eschatological transformation we refer to as the new heavens and the new earth. Predation can be influenced to some extent by human dominion but it can never finally be overcome. Being a vegetarian, for instance, requires a great deal of knowledge about food and vitamin B12, especially where children are involved.[11] Most humans even in developed countries do not have access to this knowledge. Even if the knowledge were readily available, nutritional advice is often very conflicted. Two major schools of nutrition vie for allegiance— the advocates of the *paleo*-diet, including much meat, and the plant based advocates. Attempting to untangle the evidence is no easy feat even for those with knowledge and time.[12] In a court case in New Zealand in 2001, a young vegan couple, part of an Adventist sect, went underground when they thought that medical authorities were harassing them. Their baby died of pneumonia, a result of a vitamin B12 deficiency, and the couple were both sentenced to five years in prison.[13] This case is unfortunately not unique. It highlights the difficulties associated with being vegetarian in a nonvegetarian culture, being poor and uneducated, not having a culture of trusting doctors, and being unconnected to a knowledge culture. Not all humans can easily survive on vegetables.[14] If children are also allergic or intolerant of food, as increasing numbers of them are, being vegetarian can be the last straw for their harried parents.[15]

Communion

I am convinced that food is a kind of communion as well as nourishment, and its sacramental character, the nature of sharing, and the interchange that accompanies it are sometimes more important than being strictly vegetarian, even if in an ideal world we might all be. Jesus, after all, shared these foods with us and he also shared our broken lives in all their manifestations. A part of the scandal of the idea of the divinity of Jesus is that he did participate in the multiple institutions and practices that incorporated degrees of structural evil.[16] In the same way, not all people can practice the radical hospitality that Jesus demanded and not all can be pacifists, though these too are ideals.

I agree, too, that as Southgate argues, animals now on well-run farms would not exist at all were it not for farming, and much farming in New Zealand, Britain, and parts of Europe is long established and benign.[17] Although animals that are treated badly might well not wish to live, there are many others who are treated well and show every indication of wanting to live. We are at the early stages, however, in the journey toward eating less meat. Linzey argues for progressive moves toward vegetarianism. Christians should certainly make food an issue, and must find ways to resist the powers surrounding food production in a highly technological society. There is much to be gained by eating and sourcing all food locally where the bearers of food, including meat, can give an account of its rearing and its killing. There will be animals that can be farmed for milk and eggs and cheese and other uses might be made of the land freed up by fewer farm animals—and yes there are exploitations involved in this level of farming also.

On the other hand I also agree that these justifications may turn out to be self deception. In Mary Doria Russell's novel *The Sparrow* one sentient species eats another. There is an elaborate myth by which this action is justified; the novel cuts very close to the bone. It makes us wonder, as we must, whether the nonvegetarians among us are not just attempting to justify practices that we should now resist.[18]

Communal moral imaginations and sensitivities, however, must be gradually honed; the cases against slavery and for women's suffrage similarly took time. Although the vegetarian can become the family burden, they may also be the voice of conscience that will inevitably lead us all toward a more humane future. As humans share more of their lives with animals and as the truth behind animal farming is made clearer more people will inevitably begin the journey toward a vegetarian future.

Not only is the vegetarian the voice of conscience, but perhaps the most explicit proclamation of the person who claims that at its very core the long honed practices of nature are not all good, that we live in a wheat and tares world. The vegetarian very explicitly says to the world that God does not intend the eating of flesh or the murder of one species by another, a truth hinted at by the injunction in Gen 2:29 that seeds and plants will be food for the animals of all kinds. Stopping short of claiming this is a mandatory position for all Christians is an acknowledgement of the grace that covers all our decisions and interactions in this time when the "whole creation groans," and is an acknowledgement also of the fact that we alone will not bring about the kingdom of God, much as we must work for its coming and herald its presence.

Gardens, Dominion, and Animals

Humans have learned to tame the natural world, to make gardens of great beauty. There is nothing more essentially human and yet also essentially natural than a garden. At the same time the wilderness must also be preserved and a different kind of interaction with nature takes place outside the taming of the garden. Perhaps similarly with animals; humans can learn to tame some animals, and to let others go free. These wilderness spaces are under increasing threat of exploitation and environmental degradation. The care and guardianship of these spaces is also the care of the wild animals that live in them. The tamed and domesticated animals give us relationships of great beauty, relationships that model the unconditional love of God. In a world that took animals seriously every animal would be a wanted animal, but at the same time wilderness habitats would be actively preserved and maintained and guarded.

All of this could be justified because humans have been given dominion. Such a sense of dominion mediates between the extreme ecocentrism, which sees all human intervention as unnatural and wrong—the view that sees humans as mammalian weeds—and an anthropocentrism that looks upon nature only as it is of use to humans. This discussion is very relevant to the case and controversy surrounding the polar bear Knut, born at the Berlin Zoo on March 24, 2007. Knut's mother abandoned him, admittedly under conditions of captivity. Some eco-centric activists advocated letting the cub die, saying humans had no right to interfere, even though the abandonment had happened in a zoo. Others thought the zookeeper should intervene because humans have rights over animals. The zookeeper took over the rearing of Knut in the midst of great controversy, some of it not at all straightforward.

A theocentric point of view, however, allows both that animals might not be in an ideal state just because they are a part of nature—for nature is full of tares—and that humans may be able to help because we are related in so many ways. Thus a point of view that allows for dominion might also think that any intervention that gave the young animal life was good, even allowing that human intervention at all levels can sometimes be ignorant and misguided. Deep seated beliefs about the goodness or fallenness of nature and about human prerogatives make for very different decisions in these cases.

On the other hand a theocentric position is not an anthropocentric one. Dominion and intervention can only count as a proper human exercise of dominion if it is in fact done to correct what is seen as a defect in nature—the abandonment of a cub by its mother—and if the intervention is done to give the cub a chance at life. Dominion might also be of the mutually beneficial kind, as is the relationships of having a companion animal in many cases where both "owner" and animal have a level of satisfaction not found in many human connections.

There are many other instances of human intervention or even of attempts to domesticate animals that appear to be done for the sake of humans or their curiosity—taking gorillas into human homes where they can only be controlled by drugs might count as an example of this, or keeping animals in restricted situations purely for the entertainment of humans. A Christian worldview might open up the opportunity for dominion, but it must always be exercised aware of the difficulty in really knowing any other species in its internal mindscape, and aware also of the human propensity for self-deception. However, much that we now know about animals gives us reason to believe that humans do share many human emotions and that these are also expressed in ways we understand. A person sensitive to animals will therefore have some idea of whether they are flourishing and happy in the encounter with humans. This was the case with Mark Bittner when he later watched "his" parrots being exploited by crowds in the San Francisco parks.

The Dangers of Dominion

Most animals, however, are wild and human control of their environment, and human intervention should be kept to a minimum or should be attempted only on the animals' terms. Mark Bittner's story, made famous in a book and movie, might lead one to think that interfering with and feeding wild animals is to be desired. After Mark's book and the documentary, people started feeding wild flocks in San Francisco's parks. This situation quickly got out of control. The parrots were in a feeding frenzy and people were in danger of being hurt. Moreover, poachers were exploiting this situation to steal wild birds. Mark reasoned that his own interaction had happened over a six-year period in which he had devoted almost all his time to interacting with the parrots and knowing each one by name. He came to the realization that casual interaction was good neither for the animals nor for the human feeders. Mark was responsible for the passing of a law that forbade feeding in public parks. We readily see that the story of Mark and his parrots is not without controversy. Mark himself learned through experience the limits and boundaries of human/wild animal interaction and learned that there is very little real leeway. "One thing I've learned from my experience," says Mark, "is that human beings should never enter the natural world casually or carelessly."[19] He continues:

> I was becoming more and more inclined to let nature do its own thing. I'd seen that sometimes human beings can help nature, but that more often, because we lack the proper wisdom, we create unintended, negative consequences. Often the smart thing is to back off and let nature take its own temperature.[20]

Ironically this saga was occurring in San Francisco, whose patron saint is the great tamer of wild animals, Saint Francis. Mark, however, asks the question, "What is a saint? A saint is someone who has mastered himself and has overcome his selfishness and self-indulgence. A saint is acutely aware of the effects he creates, and is responsible for everything that he does."[21] Humans can do good and can intervene for good, but in the case of wild animals the intervention must always be measured. The animals maintain an element of otherness and wildness that should not be controlled.

Naming the Animals

All of these aspects of the natural order—having dominion, and the fallen yet cooperative nature of the evolutionary process—have relevance to "naming" nature. Naming might also be integrally related to our capacities, growth, and flourishing as humans. Perhaps also the skills of both science and religion are required to "see" nature accurately. This may well be entailed, if not directly argued in Carol Kaesuk Yoon's book, *The Clash Between Instinct and Science,* mentioned in Chapter 8. She argues that it is basic to human existence to name and observe the created world; she is bewailing the lost art of taxonomy, which is, she claims, a universal, deep-seated and fundamental human activity, one we cannot afford to lose because it is essential to understanding the living world, and our place in it.[22] This understanding of the natural world common to all peoples everywhere is what she calls the *umwelt*.[23]

Yoon comes from a scientific family, is herself trained in science, and married a scientist. She set out to write quite a different book about the benefits of scientific taxonomy until two pieces of information intervened. First she came face to face with the death of the "fish." Cladists, or scientific taxonomy, are now so concerned only with evolutionary links and with DNA and chemistry that fish have been eliminated as a category; cladists take great delight in telling people that the lung fish is closer to the cow than it is to the salmon in evolutionary terms. Yet people everywhere have a category for fish, and their *umwelt* leads them to know there is a close association between the salmon and the lungfish that does not exist now with the cow. She found "there were serious clashes between the cutting edge of science and what would seem to be simple reality."[24] Thus she says that scientific taxonomy is fighting the *umwelt* and hence our current situation where ordinary human people have given up their naming rights to the scientific elite. This fighting of the *umwelt* is a part of why science has difficulty getting traction in the popular world, though this in turn leads to the lack of confidence in the *umwelt*, and a lack of attention to nature.

The second piece of information she found is that the human brain processes natural living things and inanimate objects in different parts of the brain. There are people with strokes who can still do the latter but not the former. The natural living world—and also the world of food—becomes mysterious and uncategorized to these stroke victims. Yoon thinks this dual processing is interesting and evidence that the skill for interacting with the natural living world in all its emergent levels is hard-wired. She suspects that naming of the natural world is an inherently human function that needs to be exercised; the complete lack of that ability is evident in these very sick people. Their disorientation is shared by increasing numbers of people who are shut off from nature or are indifferent to it, or those of us who do not pay attention to nature seriously and do not attempt to name it. Yoon thinks this is why we are so disconnected from the living world that we can live in the midst of "a mass extinction that worries us hardly at all."[25] "The living world is, every minute, right before our eyes, and we are missing it all."[26] Animals are only a part of nature but they are the most alive, relational and intense parts of the living world that is asking to be named. Animals often reward us in our naming of them by bonds of shared affection. This naming was discovered in all its power by the autistic

Dale's parents, and by Mark Bittner who named his wild parrots. "Whatever language we speak, we have nearly lost the language of life," Yoon warns.[27] She reflects on how odd this would have seemed to Linnaeus. "How absurd this would all have seemed to Linnaeus," she says:

> that to uncover the truth of the ordering of life, one need not know anything about animals or plants, anything of what swam or breathed or flew, of what sprouted or flowered, anything about the living world at all. One need only know about molecules, about DNA.[28]

This line of argument is interesting in light of the biblical Adam's mandate to name the animals—something Yoon also has noted; perhaps this attending to nature really is a task that is essentially religious and essentially scientific and makes us human. The encounter with nature involves confronting nature, taking notice, and then naming the living world. Perhaps both religion and science draw upon a deep human need to attend to nature, especially to the animals.

Seeing the Face of God

Behind the naming may be an even more mystical interaction with God that we have been missing, especially as we seek out God through Word alone, and in an abandonment of nature, or in a purely utilitarian approach to nature. Nature is not God, and God does not just dwell in nature, but nature may be the face of God and may be facing God in a way that is spiritually important for us as well as other creatures. Philosopher Bruce Foltz has said:

> The idea of a universe that is self-subsistent—standing entirely on its own, fully operational and intelligible, independent of anything outside itself—is both odd and modern. In the course of human experience it is an extraordinary concept, defying the shared wisdom of virtually all peoples, almost everywhere outside of Western Europe and its sphere of influence. And even within this orbit it is distinctively, and in its fully articulated configuration, exclusively modern.[29]

Nature without a face, without an interiority, self-subsistent, standing on its own—all of this is a description of the universe without God, which the new atheists so proclaim. It is a clever universe, to be sure, but not a God-filled universe. That is why their arguments are so hard to refute; a counter argument requires one to see if one has "eyes to see" that nature is alive and dependent and "given." I have argued that this discernment of otherness and givenness is particularly difficult if one has to take the evil of nature into account. Then the mysterious deeply discerned qualities of nature are easily undermined. Foltz among others would invite us to regain this experience. I argue that this is possible if a wheat and tares universe is what we might expect from the beginning. If wheat and tares are the lenses through which we come to nature we might then "see" the wheat again. He continues:

> In the Psalter, that classic anthology of Hebrew poetry and song, is to be found an aesthetic appreciation of the inner life...nature presents a face here, expresses an inner life, only because it is at the same time disclosed as being turned radically and ecstatically toward a distance unto which all the resonance of that life is directed, and from which that life is itself derived.[30]

Foltz is saying that to be alive is to bear the imprint of God. Animals, of course, speak most potently of the inwardness that is similar to us but also other, which has its own ecstatic turning toward the source of life. If Foltz and Yoon and others are right then nature is our lost reserve of spirituality and of knowledge about God. While scientific knowledge gives us much to think about, the deep knowing that so characterizes us as humans is found only in a more profound engagement with nature, akin to that of music. In the life of animals we are able to sense the traces of divinity in creation, albeit in broken ways; animals may yet take us to a deeper point of faith, they may loosen our xenophobic and species-centric doctrine. In animals it is possible that humans will discover the task of being human, and will sense if not know that the God of love is present in all of this suffering, and that God's presence is evident everywhere, in the symbiosis and the cooperation and the emerging love and pain of other sentient creatures.

Conclusion

In the end the problem of evil, especially animal suffering, forces upon us a slow conversion of perspective. Evil must be reconsidered in our theological paradigms. New metaphors must be enlivened and imagined to confront new and old evils. I have argued that seeing the world through the lenses of the wheat and the tares brings into perspective both the glory of God in nature, its perfections, but also its darker side. I argue that the wheat and the tares have characterized all life from the beginning of the story, and that humans have only enormously exacerbated that evil. Because the eschatological hope is one of peace between humans and animals and between all levels and types of life, humans have a role to play in bringing this kingdom of God to earth, of announcing a new way of being in the world in which nature matters and animals are companions along the way. I argue that new metaphors must be found not only in theology in which one of the old kernels of systematic theology has been deconstructed, but also in biology, which has fixed upon us a picture of selfish competition to the death. The grammar of this as a lone driver of evolution never was compatible with Jesus's ethic of love. Newer models of cooperation and symbiosis aid in our new way of seeing, seeing the hidden glory of God in all the splendor of life outside of the human sphere. This conversion and re-imagination of the place of humans in the world should lead especially to more care and attention and concern for God's other creatures.

Epilogue

I BEGAN THIS journey in theodicy or anti-theodicy convinced that God cannot be completely obscured within the revelation of the natural world, however troubling and difficult evolutionary history might be. I reflect on the way in which the story of evolutionary history has deconstructed some older fixed readings of Genesis 1–3. Although newer investigations of Genesis and much of the rest of Scripture open up rich alternative readings, the old story of paradise fall and renewal has been at the heart of the grammar of systematic theology for most of Christian history. It has also been at the heart of theodicy. All evil, both moral and natural, was once placed at the feet of humankind. Ignored in this monolithic theodicy was the long, long history of struggle and suffering of sentient animals. Also overlooked was the grammar of traditional understandings of evolutionary theory, as comprising an impersonal, random, law-like process that involved a highly competitive struggle that was "nature red in tooth and claw." When evolutionary theory thus understood is placed in dialogue with traditional theology we reach an impasse: why did God do it this way, only to send God's son to show us that love and turning the other cheek are the essence of moral perfection?

Similarly when contemporary people come to indwell the evolutionary story they often find that their appreciation of the God-derived nature of creation is undermined; sometimes their enjoyment of nature and their appreciation of beauty is also overwhelmed. The story we inhabit can influence greatly how we see and indwell nature. Very often this story appears to be so much in conflict with religious understandings of nature that believing in God becomes well-nigh impossible.

Recent literature has been rich in alternative offerings, in different ways of accounting for these anomalies and for the new depth that has emerged in the problem of evil. I have chartered some of these alternatives, both those within the limits of orthodoxy, and those outside it. Many twentieth century solutions were associated with or tending toward a process view of God as loving lure to the world. I argue that one gives up too much of the biblical revelation and too much of religious experience in a process view, however rich and metaphysical it might be in other ways. There are many others attempting to navigate these waters, with concepts like kenosis, proleptic eschatology, sacrifice, intratrinitarian kenosis and the pain of selving, all of which throw some light on these troubled matters; all help to some extent in the reconstruction of theology in light of evolutionary knowledge. In some sense they all tend toward the idea that this is the best of all possible worlds, and that the strong values of sentience outweigh the pain of the means to getting there, as long as there is some heavenly compensation for all animals. Against some of these positions, however, I argue that real disvalue and real evil penetrate further into created reality than we once supposed. Tares are tares wherever they reside and in spite of the difficulty in discerning them in the field of nature.

Another strong tendency in twentieth century theology and science was the idea that we cannot read God *from* the world or the process of evolution. We must believe that god has purposes *for* the created world. In light of the ongoing atheistic challenge to Christian faith I argue that this is not enough. We cannot affirm the proleptic inbreaking of the kingdom, the "deeper than Darwin," the ongoing providence of God, the indwelling Spirit, and the law of love without seeing as well as believing in nature. Otherwise we are not in good faith when we argue with the likes of Dawkins that belief is better than unbelief. This seeing means trying to look closely at nature and its processes. In the end it means affirming and discovering newer understandings of evolutionary theory as well as newer constructions in theology. These newer biological constraints and principles change the landscape of evolutionary theory for ever. They don't obliterate the specter of nature red in tooth and claw or evolutionary animal suffering, but they do paint a dualistic picture of love and hate deep in the biological substratum.

For this reason I argue that the wheat and the tares best characterize the overall process of evolution on earth. There is an almost irrational paradoxical mix of wheat and tares, of wheat depending on tares at the heart of created reality. But tares are tares nevertheless. They become enormously exacerbated under the near divine powers of human beings. Tares are perhaps visible in the slow tearing apart of an animal by its predator or the gas ovens of Auschwitz. Faced with the extent of pain and suffering and death in evolutionary history, one needs more than the lure of God to overcome these tares. Hence I link *Christus Victor* understandings of atonement to the presence of these tares and the need for a redemptive process that spans all of life and transforms reality radically, and which can and must involve human participation.

This book, then, follows the metaphor of the wheat and the tares and begins to shed light on the way in which biological and theological paradigms must be understood in its light. Behind it all is the elusive God of love, whose love might sear but never destroy, and a God whose own being is linked in covenant and incarnation to the flesh we carry and indwell and which through Christ reveals the inner grammar of the gospels, that love will endure over weakness, that weakness will overcome strength, that the poor will inherit the kingdom of God.

Human dominion in light of the wheat and the tares is a troubling and difficult task. Clearly we are to embrace our embeddedness in the process and in the ecological balance. Humans must take responsibility as the eyes and consciousness of the creation to participate in gospel grammars and not in the grammar of pillage and destruction. I ask in the end whether this means we should be vegetarians, whether we should enter into deeper relationships with animals, to what extent we can anticipate the kingdom of God. I conclude that we should at least be working toward the elimination of tares even if we can never fully identify them and cannot sustain life without them. We should also be embracing the revelatory character of a wheat and tares world in which the glory of God is evident in nature and in animal companionship, and indeed in the inner workings of the evolutionary process.

In the end theodicy is translated into struggle. God with us struggles against the tares. God with us embraces love. We in turn must embrace our troubled history, but only to redeem its goodness and not to magnify its weakness. The struggle may involve the weapons of the spirit at every level of existence being brought to bear on spiritual and embodied phenomena antagonistic to God and to goodness.

Notes

INTRODUCTION

1. J. Moltmann (1974) trans., M. Kohl, *The Crucified God: The Cross of Christ as Foundation and Criticism of Christian Theology* (London: SCM).
2. N.T. Wright (1998) "Jesus and the Identity of God." http://www.ntwrightpage.com/Wright_JIG.htm, accessed April 25, 2010, originally published *Ex Auditu* 1998, 14, 42–56.
3. M. Welker (1999) trans., J. F. Hoffmeyer, *Creation and Reality* (Minneapolis: Augsburg), p. 6 (my emphasis).
4. Welker, *Creation and Reality*, pp. 10–11 (emphasis original).
5. J. Calvin (1559/1960) ed., J. T. McNeil, trans., F. L. Battles, "The Knowledge of God Naturally Implanted in the Human Mind," *Calvin's Institutes of the Christian Religion* (Louisville: Westminster /John Knox Press), p. 43.
6. McGilchrist argues that the right brain in most people deals with and registers novelty and first impressions and living things. See Iain McGilchrist (2010) *The Master and His Emissary: The Divided Brain and the Making of the Western World* (New Haven: Yale University Press).
7. I have looked at the questions of whether animals have a shadow image in N. Hoggard Creegan (2007) "On Being an Animal and Being Made in the Image of God," *Colloquium*, 39/2, 186–204 and the salvation of animals in "Salvation of Creatures" in I. Davidson and M. Rae eds, (2011) *God of Salvation: Essays in Systematic Theology* (Farnham, Surrey: Ashgate), pp. 77–88.
8. "The wolf shall live with the lamb, the leopard shall lie down with the kid, the calf and the lion and the fatling together, and a little child shall lead them. The cow and the bear shall graze, their young shall lie down together; and the lion shall eat straw like the ox. The nursing child shall play over the hole of the asp, and the weaned child shall put its hand on the adder's den. They will not hurt or destroy on all my holy mountain; for the earth will be full of the knowledge

of the Lord as the waters cover the sea." (Isaiah11:6–9); "The wolf and the lamb shall feed together, the lion shall eat straw like the ox." (Isaiah 65:25a). This and all other Bible references are from the *New Revised Standard Version* of the Bible.

9. S. Conway Morris (2005) *Life's Purpose: Inevitable Humans in a Lonely Universe* (Cambridge: Cambridge University Press); I. Stewart (1998) *Life's Other Secret: the New Mathematics of the Living World* (New York: Wiley); P. Ball (2009) *Flow: Nature's Patterns: A Tapestry in Three Parts* (Oxford: Oxford University Press); J. Klapwijk (2008) trans., Harry Cook, *Purpose in the Living World? Creation and Emergent Evolution* (Cambridge: Cambridge University Press).

10. C. Southgate (2008) *The Groaning of Creation: Evolution and the Problem of Evil* (Louisville/London:Westminster/John Knox).

11. C. Deane-Drummond (2009) *Christ and Evolution: Wonder and Wisdom* (London: SCM).

12. In classic or *Christus Victor* atonement theories Jesus' death overcomes evil by achieving victory over the devil. In early church versions this sometimes involved God deceiving the devil into allowing Jesus to be crucified. In more contemporary Mennonite versions of *Christus Victor* Jesus' whole life is lived in resistance of powers and principalities and his death does not effect salvation but is the inevitable result of a pure life lived exposing evil. See J. D. Weaver (2001) *The Nonviolent Atonement* (Grand Rapids: Eerdmans).

13. See S. Coakley (2012) *Sacrifice Regained: Reconsidering the Rationality of Christian Belief* (Cambridge: Cambridge University Press); S. Coakley (2009) "Providence and the Evolutionary Phenomenon of 'Cooperation': A Systematic Proposal" in F. Aran Murphy & P. G. Zielger, eds, *Providence of God* (Edinburgh: T&T Clark), pp. 179–93.

14. J. Haught (2003) *Deeper Than Darwin: The Prospect for Religion in the Age of Evolution* (New York: Westview); A. Linzey *(1998) Animal Gospel: Christian Faith as Though Animals Mattered* (London: Hodder & Stoughton); A Linzey & D. Yamamoto, eds, (1998) *Animals on the Agenda: Questions about Animals for Theology and Ethics* (London: SCM); A. Linzey (2009) *Why Animal Suffering Matters: Philosophy, Theology and Practical Ethics* (Oxford: Oxford University Press); A. Linzey (2007/2009) *Creatures of the Same God: Explorations in Animal Suffering* (New York: Lantern); R.J. Russell (2008) *Cosmology From Alpha to Omega: The Creative Mutual Interaction of Theology and Science* (Minneapolis, MN: Fortress); J.C. Polkinghorne (2011) *Science and Religion in Search of Truth* (New Haven: Yale University Press); H. Rolston III (2006) *Science and Religion: A Critical Survey* (West Conshohocken, PA: Templeton Foundation Press/previous editions, 1987, 1989, 1997).

15. J. Haught (2008) "Evolution and the Suffering of Sentient Life after Darwin," in G. Bennett, M. K. Hewlett, T. Peters, R. Russell, eds, *The Evolution of Evil* (Göttingen: Vanderhoeck & Ruprecht), pp. 189–203.

16. R. Russell (2008) "The Groaning of Creation: Does God Suffer with all Life?," in Bennett, *The Evolution of Evil*, pp. 120–42.
17. Others writing in this area include H. Rolston III (1999) *Genes, Genesis and God: Values and their Origins in Natural and Human History* (Cambridge: Cambridge University Press); R. Page (1996) *God and the Web of Creation* (London: SCM); T. Peters & M. J. Hewlett (2003) *Evolution from Creation to New Creation: Conflict, Conversation, and Convergence* (Nashville: Abingdon); D. Edwards (2004) *Breath of Life: A Theology of the Holy Spirit* (New York: Orbis).
18. J. Goodall (1990/2000) *Through a Window: My Thirty Years with the Chimpanzees of Gombe* (New York: Houghton Mifflin); I. Tattersall (1998) *Becoming Human: Evolution and Human Uniqueness* (Orlando: Harcourt); F. de Waal (2005) *Our Inner Ape: The Best and Worst of Human Nature* (London: Granta).
19. J. Calvin, *Calvin's Institutes*, p. 43.
20. Jonathan Edwards argues that patterns in nature are a lower form of love. See J. Edwards (1989) ed., P. Ramsey, "The Nature of True Virtue," in *Jonathan Edwards: The Ethical Writings, volume 8* of J.E. Smith, ed., *The Works of Jonathan Edwards* (New Haven: Yale University Press), p. 564.
21. F. Schleiermacher (1989) trans., H.R. Mackintosh & J.S. Stewart, *The Christian Faith*, (Edinburgh: T & T Clark), §11.3.

CHAPTER 1

1. Arguably, this is the case even for progressive Christians who define themselves against a Christianity that emphasizes resurrection.
2. Peter Harrison says, "The narrative of the fall has always exercised a particular fascination over Western minds." It has been described in recent times as "the anthropological myth *par excellence*," "the most elemental of myths," and "the central myth of Western culture." See P. Harrison (2006) *The Fall of Man and the Foundations of Science* (Cambridge: Cambridge University Press), p. 2.
3. Karen Armstrong describes the concepts of *mythos*—timeless truth about origins and meaning—compared with *logos*, the more pragmatic and rational form of truth in K. Armstrong (2001) *The Battle for God* (New York: Knopf), p. xvi.
4. S. Grenz (1998) *What Christians Really Believe—and Why* (Cumbria: Paternoster).
5. Grenz, *What Christians Believe*, pp. 42–43 (emphasis mine).
6. R. J. Berry and T. A. Noble (2009) eds, *Darwin, Creation and the Fall* (Nottingham: Apollos).
7. D. Hart (2005) *The Doors of the Sea* (Grand Rapids: Eerdmans), p. 69.
8. J. Goldingay (2003) *Old Testament Theology, Vol 1: Israel's Gospel* (Downers Grove, IL: IVP), p. 131.
9. J. Goldingay (2010) *Genesis for Everyone: Genesis Chapters 1–16* (Louisville, KY: Westminster/John Knox), p. 61.

10. J. Goldingay, *Old Testament Theology*, pp. 131–47.
11. T. F. Torrance (1981) *Divine and Contingent Order* (Oxford: Oxford University Press); N. Messer (2009) "Natural Evil after Darwin" in R. J. Berry & M. Northcott, eds, *Theology after Darwin* (Carlisle: Paternoster), pp. 139–154.
12. K. Barth (1960) G. Bromiley ed., *The Doctrine of Creation: The Creator and His Creature, Church Dogmatics Vol. III.3* (Edinburgh: T&T Clark), p. 150.
13. K. Barth (1956) *The Doctrine of Reconciliation: Jesus Christ the True Witness: Church Dogmatics Vol., IV.1* (Edinburgh: T&T Clark), p. 479.
14. Barth, *Doctrine of Creation*, p. 42.
15. Barth, *Doctrine of Reconciliation*, p. 508.
16. See I. G. Barbour (1997) *Religion and Science* (New York: HarperCollins), pp. 84–89.
17. See J. W. Van Huyssteen (2006) *Alone in the World?: Human Uniqueness in Science and Theology* (Grand Rapids: Eerdmans). Van Huyssteen gives the example of transversal dialogue and his own book is an exemplary delving into the mysteries of human uniqueness and *imago Dei* in both theology and paleontology.
18. K. Barth (1963), trans., G.T. Thompson, *Doctrine of the Word of God: Prolegomena to Church Dogmatics, Church Dogmatics 1/1*, (Edinburgh: T&T Clark), p. 396.
19. H. Berkhof (1979) *Christian Faith* (Grand Rapids: Eerdmans), p. 49 (my emphasis).
20. J. Moltmann (2003) *Science and Wisdom* (London: SCM), p. 69.
21. W. Brueggemann (1982) *Interpretation: Genesis* (Atlanta: John Knox Press), p. 26.
22. Brueggemann, *Genesis*, p. 41.
23. Brueggemann, *Genesis*, pp. 20–21.
24. P. Ricoeur (1967) trans. E. Buchanan, *The Symbolism of Evil* (New York: Harper & Row), p. 233.
25. Ricoeur, 239.
26. Paul says in Romans 5:12 "Therefore, just as sin came into the world through one man, and death came through sin, and so death spread to all because all have sinned." This argues that Jesus is the parallel of Adam, the "one man" who brought sin into the world. This one verse has caused much angst for conservative scholars who wish to engage an evolutionary perspective on theology, and consider Adam to be anything but historical and individual.
27. A. LaCocque (2006) *The Trial of Innocence: Adam, Eve and the Yahwist* (Eugene: Wipf & Stock), p. 16 (my emphasis).
28. H. Blocher (2009) "The Theology of the Fall and the Origins of Evil" in Berry & Noble, *Darwin, Creation and the Fall*, pp. 149–72, p.157.
29. H. Blocher (1997) *Original Sin: Illuminating the Riddle* (Leicester: Apollos), p. 32.
30. Blocher's weakness is that he is more sceptical of scientific claims—though he does at least consider them—than perhaps he should be, and his strong reformed heritage leads him to consider what is probably counter-factual as easily as the probably true; evolutionary history and animal suffering do not register

as matters of equal weight at all with Scripture. See Blocher, "The Theology of the Fall," p. 167. In particular Blocher does not consider animal morality, nor the recent claims that chimpanzees, as de Waal says, "stand at the threshold of planned, organized, intercommunity conflict." See F. B. M. de Waal (1992) "Aggression as a Well-Integrated Part of Primate Social Relationships: A Critique of the Seville Statement" in J. Silverberg, & J. P. Gray, eds, *Aggression and Peacefulness in Humans and Other Primates* (Oxford: Oxford University Press), pp. 37–56, p. 40. For Blocher animal suffering is not all it is made out to be. Humans, he argues, have crossed some unique threshold and only the human story is of any ultimate significance.

31. The Cappadocians are the early church fathers, Basil the Great (330–79), who was bishop of Caesarea; Basil's brother Gregory of Nyssa (c. 330–95), who was bishop of Nyssa; and a close friend, Gregory of Nazianzus (329–89), who became Patriarch of Constantinople. The Nyssa brothers' older sister Makrina was also active in their theological group.
32. The Piraha of the Brazilian Amazon are an unusual example. They are content, mostly non-violent, and without significant material culture or mimetic desire. See D. L. Everett (2009) *Don't Sleep There are Snakes* (New York: Vintage).
33. LaCocque says: "There is a profound ambiguity indeed, in human science. It saves one hundred lives and kills two hundred. It brings humanity to heavenly heights, but invents the means to blow up planet earth many times over." See A. LaCocque, *The Trial of Innocence*, p. 248.
34. J. W. Van Huyssteen (2006) *Alone in the World*, p. 52.
35. LaCocque, *The Trial of Innocence*, p. 247.
36. I. Tattersall (1998) *Becoming Human*, p. 39; J. Diamond (2002) *The Rise and Fall of the Third Chimpanzee* (London: Vintage) pp. 261–6.
37. William Dembski in his latest book, argues for the position of a retroactive fall. Animals have been corrupted by the fall of Adam and Eve, even though most of them predated these first humans by many millennia. See W. Dembski (2009) *The End of Christianity: Finding a Good God in an Evil World* (Nashville: B&H).
38. A. Linzey (1994) *Animal Theology* (London: SCM), p. 50.
39. K. Bendall (2001) "Genes, the Genome and Disease," *New Scientist*, Feb 17 online at http://www.newscientist.com/article/mg16922787.200-genes-the-genome-and-disease.html?full=true, accessed April 25, 2010.
40. K. S. Pollard (2009) "What Makes Us Human?" *Scientific American*, 300, May, pp. 32–7, p. 35.
41. A dissenting view is taken by Jeremy Taylor in a book J. Taylor (2009) *Not a Chimp: the Hunt to find Genes that make us Human* (Oxford: Oxford University Press). Taylor quite rightly points out that although the DNA sequences might be very similar between humans and apes the organization of genetic material and gene expression are vastly different, especially in the brain. This is a much needed corrective to the idea that humans are nothing much more than apes,

but religious people tend to have the opposite problem not seeing the common inheritance at all.
42. Taylor, *Not a Chimp*, p. 137.
43. Graeme Finlay has shown clearly that errors in DNA happen haphazardly as viral segments insert themselves into DNA in humans and animals. Where these random insertions happen can be traced in terms of lineage and a branching family tree established. See G. Finlay (2003) "*Homo Divinus*: The Ape that Bears God's Name," *Science and Christian Belief* 15/1, 17–27.
44. The ubiquity of this phenomenon is described in Conway Morris *Life's Solution: Inevitable Humans in a Lonely Universe* (Cambridge: Cambridge University Press), p. 6.
45. J. Klapwijk (2009) *Purpose in the Living World?* (Cambridge: Cambridge University Press), p. 177.
46. Michael Murray refers to this as the fragility problem: Why is it that one part of the overall system so badly affects another? See M. J. Murray (2008) *Nature Red in Tooth and Claw* (Oxford: Oxford University Press), p. 82–3.

CHAPTER 2

1. C. Darwin (1872/1990) *The Expression of the Emotions in Man and Animals* (New York: New York University Press).
2. There are still some philosophers who would argue for a neo-Cartesian position. See Murray, *Nature Red in Tooth and Claw*, pp 41–72 for a discussion of this position.
3. De Waal, *Our Inner Ape*, p. 221.
4. De Waal, *Our Inner Ape*, p. 220.
5. De Waal, *Our Inner Ape*, p. 170.
6. De Waal, *Our Inner Ape*, p. 60–1.
7. This is the thesis of Jonathan Edwards who saw signs of God's presence and love in all the beauty and order and symmetry of the natural world. See J. Edwards, *The Nature of True Virtue in Jonathan Edwards*, p. 564.
8. Tattersall, *Becoming Human*, p. 68.
9. The whale expert, Anton Van Helden, was interviewed on New Zealand National Radio by Kim Hill, (Jan 26, 2008). He was pondering the great age of whales and their intelligence and wondering whether indeed they might have a religious sense of which we are unaware. I thought it interesting that for this expert a religious sense would be a sign of advanced evolutionary position, not evidence of degeneracy as Richard Dawkins and others might assume. Available as a podcast at RadioNZ, available at http://www.radionz.co.nz/national/programmes/saturday/20080126, accessed Jan 30, 2011.
10. Tattersall, *Becoming Human*, p. 66.
11. Tattersall, *Becoming Human*, p. 72.

12. De Waal, *Our Inner Ape*, 184–5.
13. Mirror neurons are brain cells which are activated when we watch another person or animal performing a certain action. This often initiates the same action in us. It also provokes sympathy for the other, even across the species boundary. So in a lecture where there is a slide show of yawning apes the humans will also yawn. A dog is known to have watched his owner limp with a broken leg and begin to limp himself. Mirror neurons will be activated if we watch another animal suffer and then we too will suffer with that animal. See de Waal, *Our Inner Ape*, p. 177.
14. De Waal, *Our Inner Ape*, p. 177.
15. De Waal, *Our Inner Ape*, p. 137.
16. De Waal, *Our Inner Ape*, p. 131.
17. Murray, *Nature Red in Tooth and Claw*, p. 71.
18. J. Olley (2000) "Mixed Blessings for Animals: The Contrasts of Genesis 9" in N. C. Habel & S. Wurst, eds, *The Earth Story in Genesis: The Earth Bible* (Sheffield: Sheffield Academic), pp. 130–39, p.130.
19. Midgley (1979/1995) *Beast and Man* (London & New York: Routledge), pp. xxviii–xxix.
20. N. J. Healy (2009) "Creation, Predestination and Divine Providence," in Murphy & Ziegler, *Providence of God* (London; New York: T&T Clark), pp. 208–31, p. 218.
21. Midgley, *Beast and Man*, p.18.
22. Van Huyssteen, *Alone in the World*, pp. 139–41.
23. Midgley, *Beast and Man*. p. 56.
24. This (controversial) claim is made by N. Patterson et. al. (2006) "Genetic Evidence for Complex Speciation of Humans and Chimpanzees," *Nature* 441, 29 June, 1103–8; See also Klapwijk, who says, "This leads to the delicate and almost embarrassing question whether there were ever on earth beings that we might have to consider, on the basis of their one-sided abilities, as partly human." Klapwijk, *Purpose in the Living World* p. 271.
25. Van Huyssteen, *Alone in the World*, p. 52.
26. There may indeed have been other species. A third species, *Homo floresiensis*, is thought to have lived on an island in Indonesia. Recently bone fragments point to yet another: See R. Dalton (2010) "Fossil Finger points to New Human Species," *Nature* 464, 472–73 online at http://www.nature.com/news/2010/100324/full/464472a.html, accessed April 27, 2010.
27. "For the creation waits with eager longing for the revealing of the children of God; for the creation was subjected to futility, not of its own will but by the will of the one who subjected it, in hope that the creation itself will be set free from its bondage to decay and will obtain the freedom of the glory of the children of God. We know that the whole creation has been groaning in labor pains until now." (Rom 8: 19–22)
28. LaCocque, *The Trial of Innocence*, p. 86.

29. J. R. Anderson, A. Gillies, & L. C. Lock (2010) "Pan thanatology" *Current Biology* 20/8, 349–51.
30. *Hamartia*, or missing the mark.
31. Brueggemann encourages the faithful art of imagining things to be otherwise than they are: "This act of imagination that inescapably constitutes our knowledge ... acknowledges that our life in the world is not simple or flat or thin or easy or obvious; it is, rather, laden with interpretive potential that is not exhausted at first glance." See W. Brueggmann (2001) *Testimony to Otherwise: The Witness of Elijah and Elisha* (St Louis: Chalice Press), p. 32.
32. Iain McGilchrist argues that the left-brain dominance of our civilization predisposes us to see the world in a left-brained mechanical way. McGilchrist, *The Master and His Emissary*.
33. Many writers in this area do not see animal death and predation as evil at all. Christopher Southgate does not believe animal death or predation is evil in itself but he does see the cutting short of animal lives before they have become true selves as evil. See C. Southgate, *The Groaning of Creation*, p. 41f.
34. R.M. Sapolsky (2006) "Social Cultures among Nonhuman Primates," *Current Anthropology* 47 (4), 641–56, 644–48.
35. LaCocque, *The Trial of Innocence*, p. 98.
36. LaCocque, *The Trial of Innocence*, p. 98.
37. LaCocque, *The Trial of Innocence*, p. 159.
38. LaCocque, *The Trial of Innocence*, p. 152 (emphasis original).
39. LaCocque, *The Trial of Innocence*, p. 116.
40. LaCocque, *The Trial of Innocence*, p. 155.
41. M. Polanyi (1958) *Personal Knowledge: Towards a Post-Critical Philosophy* (Chicago: Chicago University Press), p. 268.
42. In New Zealand, for instance, there has been ongoing tension and dispute over "ownership" of land because Maori (who were the original Polynesian settlers) have a very different, more corporate, less utilitarian relationship to land than white New Zealanders have had.
43. LaCocque, *The Trial of Innocence*, p. 263.
44. See Everett, *Don't Sleep, There Are Snakes*.
45. See my discussion of the evolutionary effects of sin in N. Hoggard Creegan (2010) "The Sin of an Evolved Humanity: Violence, Redemption and Biopower," in N. Darragh, ed., *A Thinker's Guide to Sin* (Auckland: Accent), pp. 133–42.

CHAPTER 3

1. A spandrel is defined the space between two arches, or an arch and an upright, or similar space, especially in cathedrals. A spandrel has a distinct pattern, but that pattern is derivative on the intentional structures around it. In evolutionary

biology, similarly, some patterns or functions are thought to be inadvertent by-products of the evolutionary process, not directly the result of adaption and selection themselves. See S.J. Gould & R. C. Lewontin (1979) "The Spandrels of San Marco and the Panglossian Paradigm: A Critique of the Adaptionist Programme," *Proceedings of the Royal Society, London B* 205, pp. 581–98.

2. M. McCord Adams (1999) *Horrendous Evil and the Goodness of God* (Ithaca/London: Cornell University Press).

3. In the philosophy of religion (and in Schleiermacher) further answers include the defense of soul making. This builds on the understanding that some suffering is not only helpful in producing people of character but may even be needed for any mature adult character. This defense also has some measure of truthfulness, but horrendous suffering, by its very nature, is often dehumanizing, many victims never reaching any kind of cohesion of character at all, let alone improved character. And there is no evidence at all that suffering improves the character of animals, or is evidence of soul making in them. We might still ask why an evolutionary process appears to maximize suffering for the weak and the fragile at the expense of the strong. Hick, the main contemporary proponent of soul-making, also acknowledges the difficulty of horrendous evil and moves to an eschatological approach to theodicy. See John Hick's Irenaean theodicy in J. Hick (2010) *Evil and the God of Love* (London: MacMillan, first ed. 1966).

4. Even so, the depth of human atrocity is still hard to reconcile with the goodness we often find in humanity as well.

5. Without God the problem of morality becomes very vexed. Not that people don't have a conscience and can't be moral without God, but there are huge difficulties in deciding whose morality is best; the ethic of Jesus or the ethic of the strong, for instance.

6. Michael Murray makes quite the opposite point when he argues that most so called natural evil which humans experience is partly the result of human actions or inaction—building too close to the water, not building to an acceptable standard and so on. See Murray, *Nature Red in Tooth and Claw*, p. 133.

7. This is a question often not raised in more philosophical analyses of evil where there is a quest for the value or the good that outweighs the evil. If we find this value does that then mean that we also have the right to assume that the ends justify the means?

8. "Are not two sparrows sold for a penny? Yet not one of them will fall to the ground unperceived by your Father." (Matt 10:29); "And why do you worry about clothing? Consider the lilies of the field, how they grow; they neither toil nor spin." (Matt 6: 28).

9. J. Bishop & K. Perszyk (2011) "The Normatively Relativised Logical Argument from Evil," *International Journal for Philosophy of Religion* 70/2, 109–26, 116.

10. Discussed in Murray, *Nature Red in Tooth and Claw*, p. 18.

11. Bishop & Perszyk, "Argument from Evil," 122.
12. See Matt 1:20; 2:13; 2:19.
13. Murray, *Nature Red in Tooth and Claw*, p. 21.
14. Murray, *Nature Red in Tooth and Claw*, pp. 41–72.
15. Murray, *Nature Red in Tooth and Claw*, pp. 73–106.
16. Interestingly, Rupert Sheldrake postulates a completely different paradigm for physical laws. He sees laws as habits of the cosmos at any particular time, habits which can and will be changed at some time in the future. See R. Sheldrake (2012) *The Science Delusion: Freeing the Spirit of Enquiry* (London: Hodder & Stoughton).
17. Murray, *Nature Red in Tooth and Claw*, p. 168.
18. K. Kilby (2006) "Evil and the Limits of Theology," http://theologyphilosophycentre.co.uk/papers/Kilby_EvilandLimits.pdf, accessed April 27, 2010.

CHAPTER 4

1. Process philosophy and theology now covers a wide range of positions; some theological formulations of process have emphasized only that God will not prevail by force but rather by love. See D. Ray Griffin (2004) *Two Great Truths: A New Synthesis of Scientific Naturalism and Christian Faith* (Louisville: Westminster/John Knox), pp. 111–2. Process also has many strengths, one of which is the insistence that entities have their own existence as a whole.
2. C. H. Pinnock (2001) *Most Moved Mover: A Theology of God's Openness* (Carlisle: Paternoster).
3. See for instance, J. Polkinghorne (2001) ed., *The Work of Love: Creation as Kenosis* (Grand Rapids: Eerdmans).
4. For a range of kenotic options see C. S. Evans (2006) ed., *Exploring Kenotic Christology: The Self-Emptying of God* (Oxford: Oxford University Press).
5. Moltmann, *Science and Wisdom*, pp. 53–59.
6. Moltmann, *Science and Wisdom*, p. 62.
7. Moltmann, *Science and Wisdom*, pp. 61–62.
8. Moltmann, *Science and Wisdom*, p. 63
9. Moltmann, *Science and Wisdom*, p. 63–64.
10. Moltmann, *Science and Wisdom*, p. 65 (my emphasis).
11. Moltmann, *Science and Wisdom*, p. 67.
12. J. Haught (2005) "The Boyle Lecture (2003): Darwin, Design and the Promise of Nature," *Science and Christian Belief*, 1/7, 5–20, 16–17. For further exploration of this idea see also Polkinghorne, ed., *The Work of Love*.
13. Haught, "The Boyle Lecture," pp. 13–14.
14. E. Johnson (2001) "The Word was made Flesh and Dwelt among Us: Jesus Research and Christian Faith" in J. D. G. Dunn & D. Donnelly eds, *Jesus: a Colloquium in the Holy Land* (New York: Continuum), p. 149.

15. Nevertheless Swinburne and others have argued that God must be partly hidden in order for there to be genuine freedom in good and evil. If God were so compelling we could not ignore God, this would be like holding a gun to our actions. We would have no choice but to do the good. R. Swinburne (2004) *The Existence of God* (Oxford: Oxford University Press), p. 268.
16. Moltmann, *Science and Wisdom*, p. 146.
17. Haught, *Deeper than Darwin*.
18. Southgate, *The Groaning*, p. 58f.
19. Southgate, *The Groaning*, p. 58.
20. J. F. Haught (2001) *Responses to 101 Questions on God and Evolution* (Mahwah: Paulist Press), p. 59; "The Boyle Lecture," 17–18.
21. Southgate, *The Groaning*, p. 58. Interestingly, something very similar is being argued by Celia Deane-Drummond, using the work of Von Balthasar in C. Deane-Drummond, *Christ and Evolution*, p. 184.
22. Southgate, *The Groaning*, p. 65.
23. Pelicans normally hatch two eggs, but have the resources to raise only one. The other is kept as an insurance chick and is nudged out of the nest by its sibling. Southgate speaks to this phenomenon in *The Groaning*, p. 46, taking up an idea by Holmes Rolston, in H. Rolston III (2001) "Kenosis and Nature" in Polkinghorne, ed., *The Work of Love*, pp. 43–65, p. 60.
24. Southgate, *The Groaning*, p. 35
25. Deane-Drummond, *Christ and Evolution*, p. 174.
26. R Russell, "The Groaning of Creation: Does God Suffer with All Life?," in Bennett, *The Evolution of Evil*, pp. 120–142, p. 134.
27. S. Coakley, "Sacrifice Regained."
28. S. Coakley, "Sacrifice Regained."
29. Haught (2006) *Is Nature Enough: Meaning and Truth in the Age of Science* (Cambridge: Cambridge University Press), p. 183.
30. J. F. Haught, *Is Nature Enough*, p. 184.
31. Haught, "Evolution and the Suffering of Sentient Life: Theodicy after Darwin," in Bennett, *The Evolution of Evil*, pp. 189–203, p. 202.
32. Haught, "Evolution and the Suffering," p. 201.
33. T. Peters (2006) *Anticipating Omega: Science, Faith and our Understanding* (Göttingen: Vandenhoeck & Ruprecht), p. 12.
34. Deane-Drummond, *Christ and Evolution*, p. 174.
35. I have discussed this issue in two other papers. N. Hoggard Creegan (2007) "Being an Animal and Being Made in the Image of God," *Colloquium* 39/2, 185–203; and (2006) "A Christian Theology of Evolution and Participation," *Zygon* 42, 499–518.
36. Deane-Drummond, *Christ and Evolution*, pp. 185f.
37. For Bulgakov, however, as Deane-Drummond notes, the "shadow sophia" emerged after the fall of humanity. See Deane-Drummond, *Christ and Evolution*, p. 186.

38. H. Van Till (1996) "Basil, Augustine and the Doctrine of Creation's Functional Integrity," *Science & Christian Belief* 8/1, 21–38.
39. Deane-Drummond, *Christ and Evolution*, p. 181.
40. See my paper on animal redemption in "Salvation of Creatures, in Davidson and Rae, *God of Salvation*."
41. Hart, *The Doors of the Sea*, pp. 60–61. (my emphasis)

CHAPTER 5

1. J. Cottingham (2005) *The Spiritual Dimension: Religion, Philosophy and Human Value* (Cambridge: Cambridge University Press), p. 29.
2. Hart, *The Doors of the Sea*, p. 61.
3. See for instance, *Global Spiral*, Metanexus, at www.metanexus.net April 4th, 2012, accessed April 4th, 2012.
4. C. Southgate (2011) "Re-reading Genesis, John and Job: A Christian Response to Darwinism," *Zygon* 46/2, 370–95, 383.
5. M. D. Russell (2007) *The Sparrow* (New York: Ballantine).
6. See D. Everett, *Don't Sleep there are Snakes*; For a description of the Masai of Africa who live a very similar lifestyle see V. Donovan (1978) *Christianity Rediscovered* (New York: Orbis). Vincent Donovan was a missionary to the Masai, and Dan Everett to the Piraha.
7. D. P. Domning & M. K. Hellwig (2006) *Original Selfishness: Original Sin and Evil in the Light Evolution* (Burlington & Hampshire: Ashgate), p. 30.
8. Domning, p. 5.
9. Domning, p. 110.
10. Southgate, "Re-reading Genesis."
11. Southgate, "Re-reading Genesis," p. 378.
12. Southgate, "Re-reading Genesis," p. 378.
13. Southgate, "Re-reading Genesis," p. 372.
14. Messer, "Natural Evil After Darwin," p. 149.
15. K. Barth, *The Doctrine of Creation*, p. 315, quoted in Southgate, "Re-reading Genesis," p. 382.
16. See chapter 9.
17. C.S. Lewis (1950/2001) *The Lion the Witch and the Wardrobe* (London: HarperCollins), p. 169.
18. Southgate asks why it matters whether everyone has the same levels of sadness. See Southgate, "Re-reading Genesis," p. 386. I would answer that the problem of evil is enormously exacerbated by the unevenness of affliction. This adds a dimension of injustice as well as perceived lack of ability or caring to the problem of God's work in the world. Humans can bear a great deal more if the circumstances are equal for everyone than they can when affliction or work are unevenly or randomly distributed.

CHAPTER 6

1. Hart, *The Doors of the Sea*, pp. 60–61. This theme is also explored in the recent movie, *The Tree of Life* (2011) by Terrence Malick. The film itself is a prolonged meditation on the problem of evil, opening with a quote from Job, "Where were you when I laid the earth's foundations?" Voiceovers create a dialogical space within the often silent narrative. A child asks, "why should I be good? You let a child die." Is it spoken to the father or to God; we don't know. The film insists, however, that there is glory all around us, if only we look. This is a wheat and tares movie.
2. J. Milbank (2004) "Forword" in J. K. A. Smith, *Introducing Radical Orthodoxy: Mapping a Post-Secular Theology* (Grand Rapids: Baker Academic), p. 17.
3. See E. Wainwright (2012) "Hear then the Parable of the Seed: Reading the Agrarian Parables of Matt 13 Ecologically," in R. Boer, M, Carden, J. Kelso, eds, *The One Who Reads May Run: Essays in Honour of Edgar W. Conrad* (London: T & T Clark), pp. 125–141. Wainwright uses the word *basileia* rather than Kingdom to defuse connotations of domination and the word "heavens" to denote the ambiguity in the text which incorporates both divinity and the place of the divine, the heavens or sky.
4. N. C. Habel & S. Wurst eds, (2000) *The Earth Bible in Genesis* (Sheffield: Sheffield Academic Press).
5. Some of the many theologians engaging an ecological perspective include: J. F. Haught (1993) *The Promise of Nature: Ecology and Cosmic Purpose* (New York: Mahwah, NJ: Paulist).; R. Page, *God and the Web of Creation*; D. Edwards (2005) *Jesus, the Wisdom of God: An Ecological Theology* (Maryknoll: Orbis Press); S. McFague (1987) *Models of God: Theology for an Ecological, Nuclear Age* (Minneapolis: Fortress); S. McFague (1993) *The Body of God: An Ecological Theology* (Minneapolis: Fortress); H. Rolston III (1991) "Respect for Life: Christians, Creation, and Environmental Ethics," *CTNS Bulletin*, Spring 11, 1–18; R. Radford Ruether (1992) *Gaia and God: An Ecofeminist Theology of Earth Healing* (San Francisco: HarperCollins).
6. N. C. Habel (2000) "Geophany" in Habel & Wurst, *The Earth Story in Genesis*, pp. 36–7.
7. Habel, "Geophany," p. 36.
8. E. Wainwright, "Hear then the Parable," p. 128.
9. J. F. Haught (2004) "Christianity and Ecology" in R. S. Gottlieb ed., *This Sacred Earth: Religion, Nature, Environment* (New York: Routledge), pp. 232–47, p. 236, quoted in E. Wainwright, "Hear then the Parable," P.130–31.
10. Wainwright, "Hear then the Parable," p. 133.
11. Wainwright, "Hear then the Parable," p. 135.
12. Wainwright, "Hear then the Parable," p. 140.
13. Wainwright, "Hear then the Parable," p.141.
14. Wainwright herself does not comment on the Evil One in the parable. There is a sense in which we may be meant to note the Evil One's presence but not dig too deeply into identity. Only God's presence matters.

15. J. Maritain (1959) *On the Philosophy of History*, ed. J. W. Evans (London: Geoffrey Bles), pp. 35f.
16. Maritain, *On the Philosophy of History*, pp. 36–37.
17. Maritain, *On the Philosophy of History*, p. 37.
18. Maritain, *On the Philosophy of History*, p. 38.
19. Maritain, *On the Philosophy of History*, p. 41.
20. Maritain, *On the Philosophy of History*, p. 42.
21. Maritain, *On the Philosophy of History*, p. 42.
22. R. Niebuhr (1974) "The Wheat and the Tares" in U. Niebuhr ed., *Justice and Mercy* (New York: Harper & Row), p. 51.
23. Niebuhr, *Wheat and the Tares*, p. 53.
24. Niebuhr, *Wheat and the Tares*, p. 54.
25. Niebuhr, *Wheat and the Tares*, p. 55.
26. Niebuhr, *Wheat and the Tares*, p. 55.
27. Niebuhr, *Wheat and the Tares*, p. 56.
28. Niebuhr, *Wheat and the Tares*, p. 57.
29. Niebuhr, *Wheat and the Tares*, p. 58.
30. See Gen 9:10.
31. Southgate, *The Groaning*, p. 67.
32. A kluge is a make-shift, inelegant, but workable engineering model. The human brain and eye and back are often given as evolutionary examples.
33. Hart, *The Doors of the Sea*, p. 88.
34. Everett, *Don't Sleep there are Snakes*.
35. See for instance. D. Quinn (1998) *Ishmael: An Adventure of Mind and Spirit* (London: Hodder & Stoughton).

CHAPTER 7

1. "A Picture held us captive, and we could not get outside it, for it lay in our language and language seemed to repeat it to us inexorably." From L. Wittgenstein (1961), trans. D.F. Pears & B. F. McGuinness, *Tractatus Logico-Philosophicus* (London: Routledge & Keagan Paul), and used by Jeremy Begbie in J. S. Begbie (2001) *Beholding the Glory: Incarnation through the Arts* (Grand Rapids: Baker), p. 142.
2. Richard Dawkins now downplays his earlier well known idea (meme) about selfish genes.
3. M. Midgley (2011) "The Selfish Metaphor," *New Scientist* 2797, 26–27, 26.
4. Midgley, *Beast and Man*, p. xxix.
5. S. Conway Morris (2008) "Evolution and Convergence: Some Wider Considerations" in S. Conway Morris, ed., *The Deep Structure of Biology: Is Convergence Sufficiently Ubiquitous to Give a Directional Signal* (West Conshohocken, PA: Templeton Press), pp. 46–67, p. 61.

6. J. Calvin, *Calvin's Institutes*, p. 43.
7. Carol Kaesuk Yoon has argued that naming and observing nature is a part of what makes us human. See C. Kaesuk Yoon (2009) *Naming Nature: The Clash Between Instinct and Science* (New York: Norton).
8. The New Atheists, Richard Dawkins, Sam Harris, Daniel Dennett, and Christopher Hitchens discuss the numinous among other things in an online discussion at http://richarddawkins.net/articles/2025, accessed May 4th, 2010.
9. Southgate, *The Groaning*, pp. 60f.
10. Dembski argues in a recent essay that the detection of the presence of "divine wisdom" in nature allows the kind of reaction to evil which convinced Job. It does not explain evil but it allows us to be assured of God's presence. In this essay Dembski draws on Kant's theodicy rather than the statistics which characterize much of his longer work. See W. A. Dembski (2008) "Making the Task of Theodicy Impossible? Intelligent Design and the Problem of Evil" in Bennett, *The Evolution of Evil*, pp. 218–33, p. 230.
11. Simon Conway Morris has spoken of the raw atheistic hatred of belief. Nothing exposes this deep antagonism more than any suggestion that the evolutionary process is not random and directionless. He says in his 2005 Boyle lecture, for instance, "Do I really have to remind you of our opponents' visceral aversion to religious thought and practice? ... Nor should we forget that the attitude of our opponents is not one of benign disdain, but a deep-seated animus." See S. Conway Morris (2005) "Darwin's Compass: How Evolution Discovers the Song of Creation" in *Faith Magazine*, Nov-Dec, online at http://www.faith.org.uk/Publications/Magazines/Nov05/Nov05DarwinsCompass.html, accessed May 4th 2010.
12. P. Ball (2009) *Branches: Nature's Patterns: A Tapestry in Three Parts* (Oxford: Oxford University Press), p. 206.
13. S. Kierkegaard (1844/1985) trans., H.V. Hong, *Philosophical Fragments* (Princeton: Princeton University Press), pp. 68f.
14. B. A. Gerrish (1982) *The Old Protestantism and the New: Essays on the Reformation Heritage* (Edinburgh: T&T Clark), pp. 133–9.
15. Gerrish, *The Old Protestantism*, p. 139.
16. Schleiermacher, *Christian Faith*, §13.
17. K. Barth (1957), eds, G. W. Bromiley & T. F. Torrance, *The Doctrine of God:The Knowledge of God: The Reality of God, Church Dogmatics Vol II.1* (Edinburgh: T&T Clark), p. 191.
18. Haught, *Deeper than Darwin*, pp. 159–60.
19. J. Edwards, *The Nature of True Virtue*, p. 564.
20. McGilchrist, *The Master and His Emissary*.
21. For an example of mutation rates being speeded up by starvation see S. M. Rosenberg & P. J. Hastings (2003) "Modulating Mutation Rates in the Wild," *Science* 300 (5624), 1382–3.

22. S. Conway Morris, "Evolution and Convergence: Some Wider Considerations," in Conway Morris, *The Deep Structure of Biology*, p. 62.
23. J. Sapp (2003) *Genesis: The Evolution of Biology* (Oxford: Oxford University Press), p. 63.
24. See note 8.
25. Haught, *Deeper than Darwin*, p. 151.
26. Haught, *Deeper than Darwin*, p. 144.
27. Haught, *Deeper than Darwin*, p. 144.
28. Haught, *Deeper than Darwin*, p. 129.
29. P. Berger (1990) *The Sacred Canopy: Elements of a Sociological Study of Religion* (New York: Anchor), p. 111.
30. H. U. Von Balthasar (1964) *Science, Religion and Christianity* (London: Burns & Oats).
31. J. H. Brooke (1991) *Science and Religion: Some Historical Perspectives* (Cambridge: Cambridge University Press), p. 143.
32. The Central Dogma is the biological assertion that the genome effects protein production and not vice versa; more technically it says that genome (DNA) codes for proteins (amino acids) via an mRNA intermediate.
33. See note 20.
34. Klapwijk, *Purpose in the Living World?*, p. 120.

CHAPTER 8

1. He says, "I would nominate as most worthy of pure awe the continuity of the tree of earthly life for 3.5 billion years, without a single microsecond of disruption." See S. J. Gould (2003) *I Have Landed: The End of a Beginning in Natural History* (New York: Turbo), p. 14.
2. Sapp, *Genesis*, p. 67.
3. Sapp, *Genesis*, p. 68.
4. Sapp, *Genesis*, p. 177.
5. Sapp, *Genesis*, p. 215–16.
6. Sapp, *Genesis*, p. 237.
7. F. Ryan (2009) *Virolution* (London: Collins), pp. 125–6.
8. Sapp, *Genesis*, p. 249.
9. J. Mortiz (2008) "Evolutionary Evil and Dawkins' Black Box: Changing the Parameters of the Problem," in Bennett, *The Evolution of Evil*, pp.143–88.
10. See R. A. Butler (2011) *Tropical Rainforests: The Understory*, http://rainforests.mongabay.com/0502.htm, accessed May 7, 2012.
11. Ryan, *Virolution*, p. 61, 126.
12. Ryan, *Virolution*, p. 63.
13. Ryan, *Virolution*, pp. 5–6.
14. Ryan, *Virolution*. P. 102.

15. Ryan, *Virolution*, pp. 232f.
16. Ryan, *Virolution*, p. 102.
17. Ryan, *Virolution*, p. 99.
18. Ryan, *Virolution*, p. 147.
19. Ryan, *Virolution*, p. 107.
20. Ball, *Branches*.
21. Ball, *Branches*, p. 1.
22. P. Ball (2009) *Shapes, Nature's Patterns: A Tapestry in Three Parts* (Oxford: Oxford University Press), p. 54 (my emphasis).
23. Ball, *Branches*, p. 15.
24. Ball, *Branches*, p.19.
25. Ball, *Branches*, p. 20.
26. Ball, *Branches*, p. 26.
27. Symmetry breaking, viscous flow, non-equilibrium states all lend their influence to these patterns, in life and in the natural environment.
28. Ball, *Branches* p. 208.
29. Ball, *Branches* p. 209 (my emphasis).
30. Ball, *Shapes*, frontispiece.
31. Ball, *Shapes*, p. 17.
32. S. Kauffmann (2005) *Investigations* (Oxford: Oxford University Press), p. 2.
33. Rupert Sheldrake argues that matter itself is conscious. See R. Sheldrake (2012) *The Science Delusion* (London: Hodder & Soughton).
34. Conway Morris, "Darwin's Compass."
35. Conway Morris, *Life's Solution*, pp. 1–21.
36. G. McGhee (2008) "Convergent Evolution: A periodic Table of Life?" in Conway Morris, *The Deep Structure of Biology*, pp 17–31, p. 19.
37. S. Conway Morris, "Evolution and Convergence," p. 49.
38. Conway Morris, *Life's Solution*, p. 5.
39. Conway Morris, *Life's Solution*, p. 2.
40. Parallel evolution is the development of a similar trait in different, not closely related species that have the same distant ancestor. Convergent evolution differs from parallel evolution in that the species do not have related ancestors.
41. Conway Morris, *Life's Solution*, p. 2.
42. Conway Morris, "Evolution and Convergence," p. 46.
43. Conway Morris, "Evolution and Convergence," p. 47.
44. Conway Morris, "Evolution and Convergence," p. 47.
45. Convergence in plants, for instance is considered in A. Trewavas (2008) "Aspects of Plant Intelligence: Convergence and Evolution" in Conway Morris, *The Deep Structure of Biology*, pp. 68–110.
46. Sapp, *Genesis*, p. 156.
47. Ball, *Shapes*, p. 281.
48. J. Taylor, *Not a Chimp*.

49. G. B. Müller (2005) "Evo–devo: extending the evolutionary synthesis," *Nature Reviews Genetics* 8, 943–9.
50. S. B. Carroll (2005) *Endless Forms Most Beautiful* (New York: Norton), p. 9.
51. Carroll, *Endless Forms*, p. 11.
52. Sapp, *Genesis*, p. 37.
53. Ball, *Shapes*, p. 283.
54. Ball, *Shapes*, p. 283.
55. An example of epigenesis may be a predisposition to diabetes in offspring of a mother who has eaten too much fat. See A. Coghlan (2009) "Fat reprograms genes linked to diabetes" in *New Scientist*, Sept, online at http://www.newscientist.com/article/dn17745-fat-reprograms-genes-linked-to-diabetes.html, accessed May 4th, 2010.
56. For books on emergence see: Philip Clayton and Paul Davies, eds, (2006) *The Re-Emergence of Emergence: The Emergentist Hypothesis from Science to Religion* (Oxford: Oxford University Press); H. Morowitz (2002) *The Emergence of Everything: How the World Became Complex* (Oxford: Oxford University Press); R. B. Laughlin (2005) *A Different Universe: Reinventing Physics from the Bottom Down* (New York: Basic Books).
57. Klapwijk, *Purpose in the Living World?*, p. 151.
58. Klapwijk, *Purpose in the Living World?*, p.106.
59. Klapwijk, *Purpose in the Living World?*, p.103.
60. Klapwijk, *Purpose in the Living World?*, p.117.
61. Klapwijk, *Purpose in the Living World?*, p.111–12.
62. Klapwijk, *Purpose in the Living World?*, p.114–15.
63. Yoon, *Naming Nature*, p. 12.
64. Klapwijk, *Purpose in the Living World?*, p. 177.
65. E. Johnson, "The Word was Made Flesh," p. 149.
66. Klapwijk, *Purpose in the Living World?*, p. 273.

Chapter 9

1. Hart, *The Doors of the Sea*, p. 62.
2. K. Tanner (2010) *Christ the Key* (Cambridge: Cambridge University Press), p. 63.
3. N. T. Wright (2006) *Evil and the Justice of God* (London: SPCK), p. 56.
4. Wright, *Evil and the Justice*, p. 19.
5. Wright, *Evil and the Justice*, p. 20.
6. Habel, "Geophany," pp. 36–37.
7. Wright, *Evil and the Justice*, p. 30.
8. Gen 37–9.
9. Wright, *Evil and the Justice*, p. 30.
10. Wright, *Evil and the Justice*, p. 43; See also Isa 52:13–53:12
11. Wright, *Evil and the Justice*, p. 39.

12. Wright, *Evil and the Justice*, p. 39.
13. Wright, *Evil and the Justice*, p. 42.
14. Wright, *Evil and the Justice*, pp. 28, 49.
15. Wright, *Evil and the Justice*, p. 49.
16. Wright, *Evil and the Justice*, p. 52.
17. J. D. Levenson (1988) *Creation and the Persistence of Evil: The Jewish Drama of Divine Omnipotence* (Princeton: Princeton University Press).
18. Levenson, *Creation and the Persistence of Evil*, p. xxiv.
19. Levenson, *Creation and the Persistence of Evil*, p. xxiv.
20. Levenson, *Creation and the Persistence of Evil*, p. 49–50.
21. Levenson, *Creation and the Persistence of Evil*, p. 105.
22. Levenson, *Creation and the Persistence of Evil*, p. 127.
23. J. Goldingay, *Old Testament Theology*, p. 134.
24. Goldingay, *Old Testament Theology*, p. 134.
25. Goldingay, *Old Testament Theology*, p. 140.
26. R. Jenson (1999) *Systematic Theology: The Works of God* (Oxford: Oxford University Press) p. 151.
27. R. Russell "The Groaning of Creation," p. 136.
28. R. Swinburne, *The Existence of God*, p. 229.
29. McGilchrist, *The Master and His Emissary*.
30. For a description of Peircian categories see A. Robinson (2010) *God and the World of Signs: Trinity, Evolution and the Metaphysical Semiotics of C.S. Peirce* (Leiden & Boston: Brill).
31. S. Conway Morris, *Life's Solution*, p. 24.
32. K. Barth, *The Doctrine of Creation*, p. 84.
33. Karen Kilby, for instance, has argued that "lying behind this almost universal feature of contemporary theodicies is the assumption that divine and created agency are and must be in a kind of competitive relationship"; Kilby argues the opposite may well be the case. "Although my mother may need to keep her distance," she says, "in order to allow me as an adult to develop fully into myself, God rather needs to keep as close as possible to allow this same development." See K. Kilby (2006) "Evil and Its Limits." http://www.notthingham.ac.uk/cotp/evilandlimits.doc accessed May 4, 2010.
34. J. Moltmann, *Science and Wisdom*, p. 69.
35. See N. Hoggard Creegan, "A Christian Theology of Evolution and Participation."
36. Schleiermacher, *The Christian Faith*, §13.
37. Richard Dawkins has famously said, "The universe that we observe has precisely the properties we should expect if there is, at bottom, no design, no purpose, no evil, no good, nothing but pitiless indifference." See R. Dawkins (1995) "God's Utility Function," *Scientific American* 273 (5), 80–85, 85.
38. T. L. Nichols (2002) "Evolution: Journey or Random Walk," *Zygon* 37(1), 193–210, 204.

Notes

CHAPTER 10

1. There are different understandings of original sin. In one meaning original sin is an inherited guilt common to all humanity as a result of Adam and Eve's historical fall, or it is seen as a covenant inclusion in condemnation. Yet another meaning understands original sin as an inherited fragility or propensity to sin as a result of a long history of human sinning—the inheritance can be genetic, cultural, spiritual or all of these.
2. Van Huyssteen, *Alone in the World?*, p. 37.
3. Much discussion of the fall in evangelical literature appears to be somewhat counterfactual, as though humans were the only creatures to have violent or aggressive or domineering behavior. The only relevant discussion in that case is whether the fall was the disobedience of one pair and how it might have been transmitted.
4. R. E. Watts (2002) "The New Exodus/ New Creational Restoration of the Image of God: A Biblical-Theological Perspective on Salvation" in J. G. Stackhouse Jr., ed., *What Does it Mean to be Saved: Broadening Evangelical Perspectives of Salvation* (Grand Rapids: Baker), pp. 15–42.
5. Blocher, *Original Sin*, p. 50.
6. Blocher, *Original Sin*, p. 50.
7. By an historical edge I mean that Genesis 3 refers in part to human becoming in relationship to the rest of nature, even if the telling of this account is symbolic and mythical. Because science also speaks of human becoming in other ways some sort of transversal dialogue can be established around the content of Genesis 3.
8. Michael Polanyi (1962) *Personal Knowledge: Towards a Post-Critical Philosophy* (Chicago: University of Chicago Press), p. 268.
9. Goldingay, *Old Testament Theology*, pp. 134–40.
10. A. Linzey (1998) "Unfinished Creation: The Moral and Theological Significance of the Fall," *Ecotheology* 4, 20–6, 22.
11. B. Hilton (2006) "Moral Disciplines" in P. Mandler ed., *Liberty and Authority in Victorian Britain* (Oxford: Oxford University Press), pp. 224–46.
12. C. S. Lewis (1940) *The Problem of Pain* (London: Centenary Press), p. 123.
13. C. S. Lewis, *The Problem of Pain*, p. 122.
14. T. F. Torrance (1981) *Divine and Contingent Order* (Oxford: Oxford University Press), p. 130.
15. Torrance, *Divine and Contingent Order*, p. 130.
16. Torrance, *Divine and Contingent Order*, pp. 130–31.
17. Torrance, *Divine and Contingent Order*, pp. 131–32.
18. A. Linzey (1988) "C. S. Lewis's Theology of Animals," *Anglican Theological Review* 80/1, 60–81, 70n.
19. Linzey, "C. S. Lewis's Theology," 72.
20. Linzey, "C. S. Lewis's Theology," 74.

21. R. Isaac (1996) "Chronology of the Fall," *Perspectives on Science and Christian Belief* 48(1), 34–42.
22. N. J. Ansell (2001) "The Call of Wisdom? The Voice of the Serpent: A Canonical Approach to the Tree of Knowledge," *Christian Scholar's Review* 31(1), 31–57, 36.
23. Ansell, "The Call of Wisdom," 52. Something similar is probably behind the notion of the "shadow sophia" in Celia Deane-Drummond's work. In narrative therapy also, there is a very powerful idea that one can be afflicted by "voices" that are in some sense disordered hypostases of the self. See R. Maisel, D, Epston, & A. Borden (2004) *Biting the Hand that Starves You: Inspiring Resistance to Anorexia/Bulimia* (New York/London: W.W Norton).
24. R. J. Berry & T. A. Noble (2009), "Epilogue: the Sea of Faith—Darwin Didn't Drain it" in Berry & Noble, *Darwin, Creation and the Fall*, pp. 197–204, p. 203.
25. See Chapter 1.
26. Falk (2009) "Theological Challenges Faced by Darwin" in Berry & Noble, *Darwin, Creation and Fall*, pp. 75–85, p. 75–76.
27. R. J. Berry (2009) "Did Darwin Dethrone Humankind?" in Berry & Noble, *Darwin, Creation and Fall*, pp 30–74, pp 63f.
28. T. A. Noble (2009) "Original Sin and the Fall: definitions and a proposal" in Berry & Noble, *Darwin, Creation and Fall*, pp. 99–129, pp. 120–123.
29. Berry, "Did Darwin Dethrone Humankind?," p.65.
30. Berry, "Did Darwin Dethrone humankind?," p.66–68.
31. H. Blocher (2009) "The Theology of the Fall and the Origins of Evil" in Berry & Noble, *Darwin, Creation and Fall*, pp.149–72, pp. 167–71.
32. R. Mortimer (2009) "Blocher, Original Sin and Evolution" in Berry & Noble, *Darwin, Creation and Fall*, pp. 173–96, p. 183.
33. J. Klapwijk, *Purpose in the Living World*, pp. 113–5.
34. J. Olley, "Mixed Blessings for Animals," in Habel & Wurst, *The Earth Story in Genesis: The Earth Bible*, pp. 130–39, p.130.
35. A. Linzey, "Unfinished Creation," 22; T. F. Torrance, *Divine and Contingent Order*, pp. 119–23; C. S. Lewis, *The Problem of Pain*, p. 122; M. Lloyd (1998) "Are Animals Fallen" in A. Linzey, *Animals on the Agenda*, pp. 147–60, p. 159.
36. Murray, *Nature Red in Tooth and Claw*, p. 74.
37. Murray, *Nature Red in Tooth and Claw*, p. 80.
38. Murray, *Nature Red in Tooth and Claw*, p. 83.
39. Murray, *Nature Red in Tooth and Claw*, p. 83 (emphasis original).
40. I am grateful to Miriam Bier, then a doctoral student at Otago, for drawing my attention to the themes and pathos and communal interconnectedness of the Book of Lamentations.
41. Murray, *Nature Red in Tooth and Claw*, p. 94.
42. Murray, *Nature Red in Tooth and Claw*, p. 98.
43. Murray, *Nature Red in Tooth and Claw*, p. 98.

44. Murray, *Nature Red in Tooth and Claw*, p. 99.
45. Murray, *Nature Red in Tooth and Claw*, p. 103.
46. M. Ruse (2008) "Purpose in a Darwinian World," in Conway Morris, *The Deep Structure of Biology*, pp. 178–94, p. 187.
47. M. Ruse, "Purpose in a Darwinian World," pp. 186–87.

CHAPTER 11

1. J. Goodall & M. Bekoff (2002) *The Ten Trusts: What We Must do to Care for the Animals We Love* (New York: HarperCollins), pp. 56–57.
2. N. Gardner (2007) *A Friend Like Henry: The Remarkable True Story of an Autistic Boy and the Dog That Unlocked His World* (London: Hodder & Stoughton).
3. This may at first sight appear to include the use of deception by the father, but every parent knows that metaphorical language use—and indeed some soft manipulation—is a part and parcel of child rearing. The parents could see the mutual affection of child and dog and in a sense gave a voice to the dog which the child could understand.
4. C. S. Lewis, *The Problem of Pain*, p. 65.
5. J. Edwards, *The Nature of True Virtue*, p. 561.
6. http://www.coyotecommunications.com/dogs/prisondogs, accessed May 10th, 2010.
7. See N. Habel (2004) *Seven Songs of Creation: Liturgies for Celebrating and Healing* (Cleveland: Pilgrim Press); A. Linzey (1999) *Animal Rites: Liturgies of Animal Care* (London: SCM).
8. The works of Temple Grandin are particular useful in this regard. Grandin is a high functioning autistic woman and a professional psychologist who has long worked closely with animals and especially farm animals. She has a particular skill in both handling animals and in reading their emotions and working out how to calm them. She has studied animals extensively and has designed humane animal killing procedures. See T. Grandin & C. Johnson (2009) *Animals Make Us Human: Creating the Best Life for Animals* (Orlando: Houghton Mifflin Harcourt).
9. Barbara Kingsolver describes how she and her family gave up eating meat after travelling past some of the battery farm stalls in the U.S.A. In her book, B. Kingsolver (2007) *Animal Vegetable Miracle* (New York: HarperCollins) Kingsolver describes how she does eat meat that she herself or her family have reared and killed on their own farm. Processed farm meat she does not eat.
10. On the whole in New Zealand farm animals are outside on green grass 365 days of the year. Recently, however, there was an attempt to introduce an intensive (18,000 cows) inside/cubicle dairying factory in the South Island. Resource Consent was not given and there was a widespread outcry.
11. I think it is interesting that Dr. Esselstyn, the most famous advocate of a plant based diet for health reasons, takes both vitamin B12 and vitamin D himself. He

admits that if our gut flora were perfect or un-degraded by modern conditions we might not need the B12 and people constantly in the sun might have sufficient vitamin D. See C. B. Esselstyn (2008) *Prevent and Reverse Heart Disease* (New York: Penguin).

12. I would also argue that one needs to know that knowledge is needed before one goes in search of it. Libraries and the internet also require that there be a culture of accessing knowledge in place, which for many of the world's inhabitants there is not. People seem to differ greatly in their needs for vitamins, in their metabolism and in various metabolic pathways which use and recycle key vitamins. They differ also in gut bacteria which are so essential for some nutrients and their absorption. Deficiencies, however, creep up very slowly, so slowly that there is no way that the person knows they need to know something about their diet—or that they are deficient enough to cause malnutrition in a breast-feeding infant. A significant minority of people also suffer from conditions such as irritable bowel syndrome. Some of the dietary solutions for these conditions include not eating starch, or not eating gluten, or eliminating FODMAPS, short chain carbohydrates found in fruit, wheat, onion, beans, lentils, and lactose. Adding a vegetarian requirement to these conditions is often unworkable. http://www.med.monash.edu.au/nutrition-dietetics/projects/project11.html, accessed Jan 12, 2012.

13. See http://www.nzherald.co.nz/nz/news/article.cfm?c_id=1&objectid=2045181, accessed Feb 20, 2011.

14. This would be argued by those who advocate a paleo-diet, but there are also case histories, like the one below of people who have tried vegetarianism and have been forced through ill health to give up. See for instance, http://voraciouseats.com/2010/11/19/a-vegan-no-more/, accessed Jan 12, 2012.

15. My own children were allergic to milk and eggs. Even at this level the juggling with the multiple institutions a child encounters is unmanageable if vegetarianism is added to the family routine. I would also refer to the numerous books or stories by parents attempting to "solve" the problem of their children's autism. This is not a rare problem. Most lead lives of quiet and harried desperation. One of the now standard interventions is to withdraw casein and gluten. Autistic children are almost all intensely picky eaters. To add a vegetarian regime to this scenario would be to push most of these parents over the edge of sanity.

16. Structural evil is defined as evil that works at a supra-individual level, or that determined by social practices or institutions or material culture limits. So greed might be an individual human vice but capitalism is that system within which greed is either made necessary or is encouraged beyond the bounds of an individual human will. John Wesley speaks similarly of "complicated wickedness." See J. Wesley (1826) "On Divine Providence," in The Works of the Rev. John Wesley, Vol 6 (New York: J.J. Harper) pp. 310 -20, 313. Available at http://books.google.com/books?id=oNFhAAAAIAAJ. Accessed Jan 12, 2012.

17. Southgate, *The Groaning*, p. 119.
18. M. Doria Russell (1996) *The Sparrow* (New York: Random House).
19. http://www.markbittner.net/writings/feeding_ordinance2.html, accessed May 19, 2010.
20. http://www.markbittner.net/writings/feeding_ordinance3.html, accessed May 19, 2010.
21. http://www.markbittner.net/writings/feeding_ordinance5.html, accessed May 19, 2010.
22. Yoon, *Naming Nature*, p. 12.
23. Yoon, *Naming Nature*, p. 16.
24. Yoon, *Naming Nature*, p. 10.
25. Yoon, *Naming Nature*, p. 283.
26. Yoon, *Naming Nature*, p. 21.
27. Yoon, *Naming Nature*, p. 272.
28. Yoon, *Naming Nature*, p. 238.
29. B. V. Foltz (2004) "Nature's Other Side: The Demise of Nature and the Phenomenology of Givenness" in B.V. Foltz & R. Frodeman, eds, *Rethinking Nature* (Bloomington: Indiana University Press) pp. 330–42, p. 330.
30. Foltz, "Nature's Other Side," in Foltz & Frodeman, *Rethinking Nature*, p. 334.

Index

Adam and Eve, 14–16, 39, 74, 98, 139–40, 150–51
Agriculture, 40, 73, 89, 91, 147
Animals
  and aggression, 20, 23–24, 30, 38, 41, 50
  changing views of, 23
  communication between, 29
  communities of, 28, 31, 36, 38, 41, 50, 158, 167
  compassionate behavior in, 24, 30, 36, 41, 157–58
  as connected to humans, 23, 28–30, 38, 161, 162–64
  and death, 35–38
  domestication of, 32, 144, 153, 156, 161, 166–67
  and dominion, 153, 154–61, 162–64, 166–68
  emotions in, 28, 30–32, 158
  Golden Rule and, 28–29
  and interdependence with humans, 5, 43, 154–70
  moral agency of, 23, 28–29, 32, 74, 157–61
  naming of, 168–70
  as pets, 154–60, 166–67
  salvation of, 7, 9, 11, 24, 83
  in Scripture, 11, 32, 86, 170, 171
  sentience of, 2, 5, 30–31
  and the story of creation, fall, and redemption, 23–24
  suffering of, 2, 4, 11, 41, 159
  violence in, 5, 23–24, 28, 30–31, 34, 41, 95, 145, 152
Anthropic coincidences, 68, 94
Atonement, 2, 9, 14–15, 65, 69, 129, 149, 175

Baboons, 38
Ball, Philip, 7, 114–16, 120–21
Barth, Karl, 8, 17–19, 76, 135
Basileia. See Kingdom of God
belief, 4, 12, 32, 48, 51, 53, 58, 61, 102, 111
  perspective of, 119, 125, 147
  See also Eyes of faith
Berkhof, Hendrikus, 19, 104, 147
Best of all possible worlds, 9, 10, 50–51, 71–81, 128, 174
Bishop, John, 48–53, 60, 83, 150
Blank slate view, 32–33
Blocher, Henri, 21, 139–40, 147–48
Bonobos, 28, 31, 44, 72, 152
Brueggeman, Walter, 19–20
Bulgakov, Sergii, 9, 67, 175

Calvin, John, 5, 12, 99
Chance, 6, 7, 10, 105
  See also Randomness
Chimpanzees, 25, 28–31, 77, 147, 152–53
  aggression in, 20, 28, 31, 37, 72, 92–95
Coakley, Sarah, 9, 64–65, 71, 80, 109, 111
Convergence, 7, 105, 117–19, 122, 156
Conway Morris, Simon, 98, 105, 117–19, 134, 152
Cooperation, 74, 94, 156, 168, 172
  in evolution, 7, 65–66, 71–72, 108–12
  as a life form, 28–29, 50
Creation, 2–3, 7, 18–20
  fallen, 139–53
  goodness of, 47
  imperfection and, 66
  Kenosis and, 59–62
  moral ambiguity of, 15–16, 19, 38, 62, 72–81, 89–94, 108, 118
  transformation of, 133–35
  as wheat and tares, 82–96
  wisdom and, 47, 67, 79, 102
Cro Magnon, 147, 155
Cruelty, 5, 12, 44, 54, 65, 138, 145, 152

Darwin, Charles, 1, 27, 94, 121, 147, 150
Darwinism, 57, 75, 97, 98, 101, 124, 152
  Central Dogma of, 57, 98, 109, 142
  theological response to, 135
  See also Hyper-Darwinism, Neo-Darwinism
Dawkins, Richard, 6, 47, 79, 80, 106–8, 110, 174
de Waal, Frans, 28, 153
Deane-Drummond, Celia, 9, 63, 66–70, 71, 78, 80
Death, 4, 13–17, 21, 23, 26
  and animals, 11, 35–38
  and Genesis, 38–41
  spiritual, 37, 147
  sting of, 26, 28, 35–36
Deeper than Darwin, 9, 61, 106–9, 115, 118, 122, 174

Deism, 2, 57–58, 68, 70, 104
Divinitatus sensum, 3–5, 12, 13, 99, 104
DNA, 23, 25–26, 61, 93, 111, 113, 155, 169–70
Dominion, 94, 139–40, 145, 156, 164, 175
  dangers of, 167
  and healing, 157
  and naming of animals, 168
  new forms of, 7, 153, 157–61, 166–67
Domning, Daryl, 74–75, 81
Dualism, 76, 96, 127–37

Eastern Orthodoxy, 22
Ecotheology, 25, 85–87
Eden, 2, 14–15
  in Barth, 17–19
  existential interpretation of, 15–16
  Genesis and, 19–22
  historical interpretation of, 15, 21
  and original sin, 75
  as paradise, 2, 15–16, 98, 139–40, 150, 154–55
  story as kernel of theology, 15
Edwards, Denis, 85
Edwards, Jonathan, 42, 72, 104, 107, 160
Embodiment, 29–30
Emergence, 7, 34, 105, 108, 116, 122–24, 134, 156
Empathy, 28, 30–33, 39
Eschatology, 9–10, 34, 64, 164, 172
Evil
  dualism and, 78, 127–37
  in evolution, 4, 7–13, 46–47, 53, 74, 80, 93, 127–37
  as fall, 138–53
  God responsible for, 48–49
  God's hiddenness and, 133–36
  gratuitous, 51, 54, 80
  horrendous, 24, 45, 49, 69, 79, 151
  inside and outside sources of, 39, 130–33
  moral, 46
  natural, 46, 82–84
  normalizing of, 63, 65

as nothingness, 76
and predation, 46–47
and resurrection, 52, 69–70, 128–30
secular responses to, 1
temptation as, 78, 130–33
See also Theodicy
Evo devo, 117–22
Evolution, 4, 14
  Central Dogma of, 57, 98, 109, 142
  constraints and, 105, 117
  convergence and, 7, 105, 117–22
  cooperation and, 7, 111–12
  as emergence, 105, 122–24
  empirical status of, 21
  and *evo devo*, 117–22
  grammar of, 21
  jumping genes and, 25, 93, 111
  and Lamarckianism, 111
  and mathematics, 65, 108, 114–16, 116, 125
  mythical significance of, 21
  necessity vs contingency and, 110
  new dynamics and, 110–26
  and niches, 98, 105, 112, 117–18
  physics and chemistry of, 105, 114–16, 126
  pictures of, 97–99
  as punctuated equilibrium, 110
  purpose in, 117–21, 124, 156
  sacrifice in, 64–64, 72, 112
  self-organization in, 116
  viruses and, 112–14
  See also Darwinism, Natural selection
Existentialism, 20, 33, 36, 38, 139, 143
Eyes of faith, 70, 87, 94, 103, 152

Fall, 1, 14–17, 21, 139–42
  and animals, 23
  arguments against, 148–51
  as corruption, 82–84
  cosmic, 142–46
  as disorder, 142–44
  and Eastern Orthodoxy, 22
  existential interpretations of, 20
  historicity and, 21
  as kernel of theology, 15–17, 42
  new arguments for, 146–48
  as progressive, 40–41, 140–42, 146
  re-imagining, 151–53
Finlay, Graeme, 25
Freedom, 33–34, 42, 134–35
  of Adam and Eve, 74
  and theodicy, 48–50, 52, 54, 59, 79, 143, 150
Fundamentalism, 4

Gardens, 89, 91, 94, 166
Genesis, 14–20
God
  beauty and, 6, 12, 45, 94, 104
  of Deism, 57, 104
  as discerned in nature, 6, 104
  hiddenness of, 102–4, 133–36
  and intratrinitarian kenosis, 9, 61–63, 131
  nature of, 2–5
  as omni-God, 47–51, 56
  of the philosophers, 2–4, 47–54, 56–58, 68–69, 79
  of popular opinion, 12
  of process theology, 57–58
  providence of, 54, 125, 129, 174
  purposes in creation, 12, 19, 58, 60, 71, 76, 80, 87, 108, 117–19, 134, 174
  of Scripture, 2–4, 48, 51
  and *Shekinah*, 59
  sovereignty of, 127, 133
  as Trinity, 18, 48, 63, 102, 134
Goldingay, John, 16, 21, 132–33, 139, 141
Gould, Stephen J., 99, 110
Greater good, 52
Grenz, Stanley, 15

Habel, Norman, 85–86, 129
Hart, David, 16, 69–70, 71–72, 78, 80, 82, 94

Haught, John, 9, 60–61, 66, 85–86, 103, 106–8
Healing, 58, 69, 128, 149, 156
  dominion and, 157–64
Healy, Nicholas, 33
Hellwig, Monika, 74–75, 81
Holocaust, 1, 46, 151
Holy Spirit, 47, 51, 54, 57, 60, 62–63, 68, 79–80, 102, 103, 126, 136, 174
Hominids, 23, 34, 53, 83, 140–41
Homo Sapiens, 26, 113, 140, 152
Human Genome, 25, 33
Human nature, 32–33
Humans
  as animals, 11, 15, 33–35
  in solidarity with all life, 43
Hyper-Darwinism, 44, 45, 97, 101, 135

Imago Dei, 15, 19, 29, 34, 35, 152, 154
  in animals, 141, 147
Incarnation, 60, 68, 76, 79, 133, 136, 175
Intelligent Design, 12, 97, 105
Isaiah, 7, 70, 102

Jesus Christ, 1, 8, 19, 32, 73, 149, 155, 164
  birth of, 79, 133
  divine/human status of, 102–3
  and ethic of love, 44, 172
  temptation of, 76, 79, 141
  See also Incarnation, Resurrection
Job, 6, 13, 81, 100–2, 108, 125

Kenosis, 12, 54, 59–61, 66, 70, 108, 174
  intratrinitarian, 9, 61–63, 131
Kierkegaard, Søren, 102
Kingdom of God, 10, 60, 66, 85–86, 92, 114, 128, 161, 166
  as immanent, 114, 156–57
Klapwijk, Jacob, 26, 34, 109, 123–25, 136, 148, 154

LaCoque, André, 20, 22–23, 35, 37–39
Lamarckianism, 111

Lamentations, 6, 37, 150
Levenson, Jon, 130–32
Lewis, C.S., 77–78, 142–45, 159
Liberalism, religious, 4
Linzey, Andrew, 10, 25, 69, 141, 144–46, 149, 164–65
Logos, 47, 62, 126

Malick, Terrence, 80
Mammals, 11, 25, 29–30, 123
Maritain, Jacques, 87–88, 90–92, 108
Mathematics, 12, 34, 65, 94, 136
  and evolution, 7, 102, 105, 107, 108, 114, 116, 125
McGilchrist, Iain, 104, 134
Meerkats, 50
Messer, William, 75–76, 78
Midgley, Mary, 32–34, 98, 125
Mimetic desire, 30, 33, 41, 152
Mirror neurons, 30, 32, 183
Moltmann, Jürgen, 1–2, 19, 59–62, 69, 80, 85, 135, 147
Moral order, 130
Morality
  in animals, 11, 20, 23, 27–29, 32, 42, 155, 181
  communal, 165
  development of, 63
  and empathy, 28, 31
  and evolution, 35, 41, 52, 74, 91, 101
  of God, 64
  hiddenness and, 133
  and mirror neurons, 30
  moral codes and, 123
  and perfection, 139, 173
  See also Proto-morality
Murray, Michael, 5, 9, 31, 45, 51–53, 78, 149–51

Natural selection, 7, 25, 44, 50, 53, 60, 74, 104–6, 112, 116, 126, 142, 147, 155
Neanderthal, 34, 154

Neo-Cartesians, 31, 45, 52, 182
Neo-Darwinism, 69, 72, 99, 107, 120, 124, 125, 142, 155
Niebuhr, Reinhold, 88–94
Nutrition, 164

Original sin, 20, 35, 66, 74–75, 139

Pannenberg, Wolfhart, 34
Parables, 6, 82–94, 128, 133
Peaceable kingdom, 37, 70, 83, 155, 157
Pelicans, 63, 72, 159
Perszyk, Ken, 48–53, 60, 83, 150
Pets, 32, 157–58
Piraha tribe, 41, 73, 95
Plantinga, Alvin, 49–51
Polkinghorne, John, 10
Powers and principalities, 8, 68, 92–93, 129–30, 134, 137, 178
Predation, 2, 6, 46–47, 77, 91, 119, 137, 140–46, 157, 162, 164, 184
  causes of, 31
  kinds of, 50, 73
  as necessary for evolution, 8–9, 93–95, 138
  and the peaceable kingdom, 37
  skills in, 24–25
  and theodicy, 53–54, 70, 84, 151–52
Primates, 23–25, 28–38, 118, 154
Proto-morality, 32, 44, 46

Randomness, 28, 44, 65, 95, 115, 123, 142, 155–56
  in evolutionary process, 97–98, 101–2, 105–8, 119–20, 125–26, 142, 173
  and intelligent design, 105
  and viral inserts, 113
Redemption. *See* Atonement
Religious experience. *See* Divinitatus sensum
Resurrection, 24, 36, 52, 67, 69–70, 128–30, 145, 179
Ricoeur, Paul, 20, 147

Russell, Mary Doria, 73, 165
Russell, Robert, 10, 63–64, 78, 133
Ryan, Frank, 112–13

Sapp, Jan, 106, 109, 110–11, 120–21
Satan, 8, 52, 93, 143, 146, 149, 151
Schleiermacher, Friedrich, 13, 42, 103, 135, 185
Science, relation to faith, 18–22
Self-consciousness. *See* Sentience
Selfish gene, 29
Selfishness, 35, 74, 90
Selving, 9, 62–63, 100, 123
Sentience
  in animals, 5, 20, 30, 32, 36, 52, 98, 123, 137
  dark side of, 40,
  degrees of, 2, 29, 44, 48, 54, 73, 78, 84, 118, 146
  and empathy, 28, 30–31
  human, 20, 30, 52, 77, 174
  mirror neurons and, 30
Shadow sophia, 9, 67, 70, 71, 77, 80, 133, 137
Shoah, 1, 4, 6
Sin, 9, 15–16, 18, 19, 20, 33, 35, 36, 49, 74, 145
  of Adam and Eve, 139
  effects of, 45, 146
  and freedom, 42
  at species boundaries, 42, 46
  unconscious, 162
  See also Original sin
Southgate, Christopher, 61–63, 72, 75–78, 93, 148–49, 165
Spirituality, 34, 79, 147
  in animals, 34, 37
  nature and, 171
Stewart, Ian, 7, 114
Suffering, 73, 175, 185
  animal, 1, 4, 52
  and evil, 11
  God's, 79

Supernatural, 50, 135–36
Symbiosis, 84, 108–13, 117, 121, 156, 159, 171–72

Tattersall, Ian, 11, 29
Temptation, 16, 40, 78, 93, 141
Theodicy, 11
  Causa Dei and, 51–53
  chaos or order and, 52–53
  deism and, 57–58
  evidential argument and, 51–53
  of free will defence, 48
  of fully gifted universe, 68–69
  of glory, 59, 70, 80, 94, 134, 136
  God of Scripture and, 48, 51
  of imperfection, 66–67
  and intratrinitarian kenosis, 61–63
  and Job, 6, 13, 81, 100–2, 108, 125
  kenosis and, 59–61
  logical argument, 47–51
  middle knowledge and, 49
  NIODA and, 10, 63–64
  nomic regularity and, 52–53, 83
  omni-God and, 47–51, 56
  original sin and, 74–75
  prayer and, 11, 49–50, 107
  of process, 57–60
  redemption and, 127, 129
  resurrection and, 69–70, 128
  sacrifice and, 64–66
  shadow sofia and, 9, 67–68, 70, 77
  of soul-making, 45, 52, 151
  speculative theories and, 53–54
  spirituality and, 79–80
  suffering of God and, 1, 25, 79
  Theology of hope and, 1
  in *The Tree of Life*, 80
  theodrama and, 67–69
  theologians and, 79–80
  See also Best of all possible worlds, Sarah Coakley

Theodrama, 67–69
Theology, 4, 32
  Eastern Orthodox, 22
  of hope, 1, 64, 70
  kernel of, 14, 15–17, 42, 154, 172
  paradox in, 55, 76, 103
  of Process, 57–60, 67, 104–5, 174
  systematic, 11, 14–26, 149, 173
Torrance, Thomas, 144–45
Tragic, 22, 26, 37, 38–40, 78, 82, 112

Van Huyssteen, J. Wentzel, 23, 34, 139, 154
Vegetarianism, 77, 152, 161–65, 175
  difficulties with, 163–64
Violence, 2, 5, 8, 16–17, 65, 79
  in animals, 24, 30–31, 41, 92–93, 95, 146–47
  in humans, 30, 41–42, 73, 95
  in nature, 69, 75–76, 91
  origins of, 21, 34, 35, 72
  and predation, 140
  primary, 65
Von Balthasar, Hans Urs, 67

Wainwright, Elaine, 86–87
Welker, Michael, 3
Wheat and Tares, 73, 80, 82–96, 113, 127–37
  and animals, 31, 92–93, 128, 161–65
  in ecology, 85–87, 91–94
  in Jacques Maritain, 87–88
  in Reinhold Niebuhr, 88–91
  theodicy of, 61, 64, 66, 68, 125, 138
  as view of the universe, 6–7, 76
  in Elaine Wainwright, 86–87
Wittgenstein, Ludwig, 97, 109
Wright, N. T. (Tom), 2–3, 77, 128–30

Yoon, Carol Kaesuk, 124, 168–71